Double or Quits?

The Global Future of Civil Nuclear Energy

Malcolm C. Grimston and Peter Beck

THE ROYAL INSTITUTE OF INTERNATIONAL AFFAIRS | Sustainable Development Programme

Earthscan Publications Ltd, London

First published in the UK in 2002 by
The Royal Institute of International Affairs, 10 St James's Square, London SW1Y 4LE
(Charity Registration No. 208223)
and
Earthscan Publications Ltd, 120 Pentonville Road, London N1 9JN

Distributed in North America by
The Brookings Institution, 1775 Massachusetts Avenue NW,
Washington, DC 20036-2188

A catalogue record for this book is available from the British Library.

ISBN   1 85383 908 6 hardback
          1 85383 913 2 paperback

The Royal Institute of International Affairs is an independent body that promotes the rigorous study of international questions and does not express opinions of its own. The opinions expressed in this publication are the responsibility of the authors.

Earthscan Publications Ltd is an editorially independent subsidiary of Kogan Page Ltd and publishes in association with the WWF-UK and the International Institute of Environment and Development.

Typeset by Denis Dalinnik
Printed and bound in the UK by Creative Print and Design (Wales), Ebbw Vale
Original cover design by Visible Edge
Cover by Yvonne Booth

Contents

List of figures and tables

Figures

Tables

Acknowledgments

This project has only been possible through the financial support of the following organizations:

- British Nuclear Fuels plc
- Cox Insurance Co.
- Department of Trade and Industry, UK
- Electricité de France
- Golden Rule Foundation, USA
- International Atomic Energy Agency
- Kansai Electric Power Group Inc.
- TotalFinaElf

Furthermore, the project has greatly benefited from the support of an Advisory Committee, which through meetings and written comments has contributed enormously to the development of the thoughts presented here. Members of that Committee have been:

Sir Crispin Tickell (Chairman)	Former Warden, Green College, Oxford; past UK representative on UN Security Council.
Ed Arthur	Deputy Director, Los Alamos National Laboratory, USA.
Sir Eric Ash	Vice President and Treasurer, Royal Society, London.
Fred Barker	Consultant (e.g. to Nuclear Free Local Authorities, Friends of the Earth).
Matthew Bunn	Kennedy School of Government, Harvard University.
Gerald Clark	Nuclear policy analyst/consultant.
Prof. A. Gagarinski	Director, International Affairs, Kurchatov Institute, Moscow.
Malcolm Keay	Former Deputy Head of Programme, Sustainable Development Programme, RIIA, London.
Stewart Kemp	Secretary, Nuclear Free Local Authorities, UK.
David Lascelles	Co-Director, Centre for the Study of Financial Innovation, London.
Gordon MacKerron	NERA Consultancy, London.

Jan Murray	Deputy Director-General, World Energy Council.
Prof. A. Suzuki	Professor of Nuclear Engineering, University of Tokyo.
John Ritch	Director-General, World Nuclear Association, London; past US Ambassador to IAEA.
Silvan Robinson	Chairman, British Biogen; former Head, Energy and Environment Programme, RIIA.
Caroline Varley	Head, Energy Diversification Unit, IEA, Paris.
Acad. Ren-kai Zhao	Senior Advisor, China National Nuclear Corporation.

We have in addition received a wealth of written responses from experts within the fields of energy, sociology and politics and held a number of well-attended international workshops – we would like to thank all involved. We are also grateful to the staff of the Sustainable Development Programme, RIIA and especially to its Head, Duncan Brack, for keeping us on track; to Kate Kinsman and Ruth Tatton-Kelly for the invaluable assistance given to us, for organizing the workshops (a thankless but essential task) and for spurring us to deliver the manuscript more or less on time; and to Margaret May, Head of RIIA Publications Department, and Dr Kim Mitchell, who laboured hard to make the book more readable. However, responsibility for the final text, and whatever errors it may contain, is ours alone.

May 2002   Malcolm Grimston
                                                              Peter Beck

About the authors

Malcolm C. Grimston is an Associate Fellow in the Sustainable Development Programme at the Royal Institute of International Affairs. After reading Natural Sciences at Magdalene College, Cambridge, he taught chemistry for seven years. In 1987 he joined the UK Atomic Energy Authority as an information officer. In 1995 he was appointed as a Senior Research Fellow at the Centre for Environmental Technology, Imperial College, London, working in the Energy and Environment Policy Group, and retains Visiting status at the college.

Malcolm Grimston's publications include *Coal as an Energy Source,* for the International Energy Agency, *Chernobyl and Bhopal Ten Years On: Comparisons and Contrasts* and *Leukaemia and Nuclear Establishments – Fifteen Years of Research,* as well as numerous journal articles. He is a regular media contributor on energy and nuclear matters.

Peter Beck is an Associate Fellow in the Sustainable Development Programme at the RIIA. He is an adviser to the Oxford Institute for Energy Studies, and has been a member of an international working group on setting up an internationally monitored storage and management regime for spent reactor fuel and plutonium from civil nuclear power plants.

Peter Beck has been active in matters relating to business planning and strategy, having been Planning Director of Shell UK Ltd during a career of nearly forty years with the Shell Group, Chairman of the Strategic Planning Society in UK and subsequently President of the European Strategic Planning Federation. He has published and lectured widely on subjects related to the future of nuclear power, the practice of strategic planning, energy policy and the futility of much of energy forecasting.

Both authors collaborated on a preliminary study for this volume, *Civil Nuclear Energy: Fuel of the Future or Relic of the Past?* (RIIA, 2000).

1 Setting the scene

The energy challenge

It is clear to many commentators that the world cannot continue along its current energy course. Among the many reasons why it cannot do so, perhaps two stand out. The first is the challenge of supplying enough energy for a world population that is not only growing but also increasing its average per capita energy use. Global energy demand is expected to rise to double or treble its present level by mid-century, while oil production from 'conventional' sources, which accounted for about 40 per cent of traded primary energy in the year 2000, may peak within two or three decades. Conventional reserves of gas will last rather longer, but by mid-century they too may be under heavy pressure, especially if gas supplants oil. Unconventional energy resources such as shale oil or methane hydrates may complement the conventional ones, but to what extent is not clear.

At the same time, there is a growing consensus among many climatologists and other scientists that emissions of greenhouse gases, of which carbon dioxide is the most important, must fall by some 60 per cent by 2050. This is essential if their concentration in the atmosphere is to be stabilized at double the pre-industrial level, allowing climate change to be restricted to manageable proportions. Nonetheless, a doubling of greenhouse gases in the atmosphere will still, by current calculations, have very significant environmental consequences.[1]

Fossil fuels provide some 90 per cent of commercially sold primary energy in the world today.[2] Of the remaining 10 per cent, nuclear power provides about three-quarters of 'input' energy and hydropower almost all of the rest. (Once corrections have been made for losses in converting heat into electricity in nuclear stations, the present contributions of nuclear power and hydropower are approximately equal.) In addition there is the use, mainly in less industrialized countries, of non-commercial energy sources such as wood and animal wastes which accounts for an extra 10 or 15 per cent, a proportion likely to fall as more people gain access to modern energy sources.

A crude calculation, assuming a doubling of global energy use by 2050, normal improvements in the efficiency of fossil fuel use and a continuing shift

---

[1] See Depledge (2002).
[2] BP Amoco (2001).

from coal to gas, suggests that fossil fuels should be providing no more than 30 per cent of global energy in 2050 to achieve the required reductions in greenhouse gas emissions. Sources which do not emit greenhouse gases (renewables and nuclear power) would have to grow by a factor of about 15 from year 2000 levels for this to happen or by a factor of 50 if nuclear power is excluded.

These rates of growth look unrealistic at present. Against this background a reappraisal of all approaches to providing energy is essential. This reappraisal must determine what actions are required in the near future to ensure that over the next decades sufficient supplies and types of energy will be available to meet both growing world demand and greenhouse gas emission constraints.

There appear to be four major options for addressing these challenges. They are:

- demand-side measures to reduce the requirements for energy;
- nuclear power;
- renewable energy sources;
- sequestration of carbon dioxide produced during the use of fossil fuels.

Although these options are sometimes seen as being in competition, this is by no means necessarily the case. Given the enormous uncertainties that lie ahead in the energy field, on both supply and demand sides, it is quite possible that some combination of all four will be needed. And while each has its attractions, each also offers its own challenges.

Measures to reduce energy use are in a different category from the other options. Improvements in energy efficiency have been a constant feature of technological development since the start of the Industrial Revolution, and the question for the future is largely about the rate at which this improvement will continue. It is also generally accepted that reducing the use of energy is the most attractive option in terms of public perception – it is inconceivable that public campaigns could develop in opposition to energy reductions, in the way that one could foresee opposition to nuclear power, renewables or carbon dioxide sequestration.

Certainly much can be done in terms of improving the efficiency of energy conversion and end-use. However, there are paradoxes in this field. The rate of take-up of improved technology and techniques has historically often been disappointing, even when the expected payback period is short. The experience of the 1970s and 1980s suggests that legislative measures can help, although they are sometimes unpopular politically.

Further, there is a vigorous debate over the extent to which improvements in energy efficiency result in reductions in energy use. As improvements in energy efficiency (taken alone) reduce the effective cost of using energy, one expected result of improved energy efficiency would be to make economic activity more attractive. This in turn would have implications for total energy use.

There is considerable dispute over the extent of these 'rebound effects'. Some commentators observe that the steady improvement in energy efficiency over the past two centuries has been accompanied by enormous increases in the total amount of energy used, and offer an analogy with the introduction of the jumbo jet. It was believed that larger aircraft would reduce the number of flights required. In reality, larger planes resulted in a fall in the unit cost of air travel, allowing more people to fly. The result was an increase in the number of flights as well as in the carrying capacity of each individual flight. For these commentators, attempts to improve energy efficiency will promote economic growth, but will be unlikely to result in major reductions in energy use.

Other commentators argue that the size of the rebound effects is likely to be small. In a modern 'post-industrial' economy in which many people live in comfortably warm conditions and do not have major needs for basic commodities, the energy implications of an effective increase in income (caused, for example, by people saving money because of measures to use energy more efficiently) are modest. Rebound effects may therefore be no greater than 10 per cent of the improvement achieved in energy efficiency, i.e. an improvement in energy efficiency of 10 per cent would result in a 9 per cent reduction in energy use.

It is not the role of this study to adjudicate between these positions. However, two observations may be appropriate. First, if the price of energy is increased to 'compensate' for improvements in energy efficiency, so that on balance people still have to pay the same amount for the energy services they consume, then the rebound effect should be neutralized and genuine reductions in energy use should be possible. But the thrust of liberalization in industrialized economies in recent years has been to reduce the costs of energy production, so allowing for price reductions in the market place. Second, even if there should be major improvements in energy efficiency, it is extremely difficult to believe that these improvements alone can come close to compensating for the enormous projected increases in global demand for energy services over the next decades.

Renewable energy sources generally do not release greenhouse gases or other noxious wastes, and are not fuelled by limited reserves that have other beneficial uses. However, there is much debate about the degree to which they could be deployed in practice. They tend to require large amounts of machinery, spread over large areas of land or sea, owing to their low energy density. This has obvious environmental consequences, both during the manufacture of the machinery and through its installation and use. Some of the renewable sources are also intermittent, either predictably (tidal power) or unpredictably (wind power). This is a serious issue in the absence of a large-scale way of storing electricity, especially if intermittent renewables were to provide more than about 20 per cent of electricity supply. There are also question marks over the costs, especially of installation, of some options, although considerable progress has been made in reducing them in recent years.

At present, direct carbon dioxide sequestration – capturing carbon dioxide from waste gases and storing it underground, in deep oceans or in the form of solid carbonates – appears to be a very costly option, adding perhaps between 40 per cent and 100 per cent to electricity generation costs. (Indirect sequestration in carbon 'sinks' such as forests is cheaper but is controversial.) There are also questions associated with how secure the final stores will be: is there a danger that carbon dioxide might escape into the atmosphere and, if so, when? Furthermore, it is not clear how efficient the process might be, especially if applied to flue gases with a relatively low proportion of carbon dioxide, such as those that come from conventional coal-fired or gas-fired power stations. The current state of knowledge and experience is too limited to make a firm assessment of sequestration's potential contribution to reducing carbon dioxide emissions.

This book acknowledges these uncertainties, but its main purpose is to focus on the benefits and disadvantages of civil nuclear power as a source of electricity. Its particular objective is to consider what actions are necessary, in practical terms, to keep the option of further investment in new nuclear reactors open, even in those countries where it looks unattractive today.

Timescales

Decision-making in the energy field is further complicated by the fact that it has to deal with contrasting timescales. On the one hand, things can change very rapidly – within a year (the world oil price trebled between mid-1999 and mid-2000), a month (the electricity shortages in California late in the year 2000 appeared quite suddenly), a week (the transportation fuel crises in much of Europe in September 2000) or even a day (11 September 2001 underlined the danger of terrorist attacks on energy installations such as nuclear stations or large power dams). However, even quite radical changes are often reversed, sometimes just as quickly.

Action is required today, or very soon, in order to make a radical difference to the energy options available in 2050. Some of the decisions taken now in the energy field can have implications for many decades, if not longer. There is often a long period between the emergence of a new concept and its availability for commercial exploitation. In many industrialized countries, a decision taken today to build a nuclear power station may well be followed by some years of planning and regulatory activity, a construction phase lasting up to six years and 50 years of operation. The installation of a new gas pipeline or a major hydropower or tidal facility may have similar time horizons. In a competitive power market in which investors generally prefer a quick return and are averse to economic risk (which inevitably increases as timescales lengthen), the time gap between the emergence of a new concept and its commercial availability is accentuated.

Timescale is perhaps an especially important issue in the case of nuclear power, for two reasons. The first is the age profile of the world's nuclear power capacity. If the average lifetime of a nuclear power reactor is taken to be 40 years, it can be seen from Table 1.1 that more than one-third of present installed capacity will come to the end of its operational life before 2020 and that a further 51 per cent of current capacity will do so between 2020 and 2030. Even if all projects currently described as 'under construction' are completed, they will add only a further 26,500 MW, giving a total of about 70,000 MW in operation in 2030 (assuming some life extension). This is a mere 20 per cent of current installed nuclear capacity.

The 'lost' 80 per cent of this capacity will have to be replaced. There should be no assumption that nuclear power will necessarily be replaced by nuclear power, but if it is not, then the replacement may have to be fossil-fired (with or without sequestration of carbon dioxide, and in either case with some implications for resource management and greenhouse gas emissions). And if it

Table 1.1 Age profile of world nuclear power capacity, 2001

Age (years)	Capacity (MW (e))	Percentage of total
0–10	41,800	11.9
10–20	179,800	51.0
20–30	117,200	33.3
30–40	13,200	3.7
40+	400	0.1
Total	352,400	100.0

Source: IAEA (2002).

is not fossil-fired, it will have to be achieved by renewables or by energy-demand reductions. In either of these cases, supporters of nuclear power would argue, an opportunity would have been lost. Instead of using renewables or demand reduction measures to replace use of fossil fuels, thereby reducing emissions of greenhouse gases, they would simply be used to replace one low-carbon source of energy services with another, with no net atmospheric benefit.

The second reason arises from the relative inflexibility and irreversibility of investment in very large plants, such as traditional nuclear stations. In competitive markets there is considerable advantage in being able to change the mix of fuels used for power generation as rapidly as possible in response to changes in market conditions. The very high initial investment necessary to build a traditional nuclear plant (or other large project) in effect ties the operator to the technology to a much greater extent than would be the case, for example, with a relatively small-scale gas-fired plant. Yet reductions in the costs of nuclear investment may be highly dependent on ordering a series of four, six or even eight plants. The trade-off between costs and flexibility is an awkward one for large plants of any description, and perhaps traditional nuclear reactors in particular. The development of modular approaches to nuclear power generation, which are based on much smaller units, may help to address this issue.

Different futures

In view of the above uncertainties and timescales, the only rational approach is to keep as many options open as possible. If the future is uncertain, we have to plan for many different futures, accepting that in hindsight some resources will appear to have been wasted.

It is thus difficult to argue against the proposition that the option of building more nuclear stations should be kept open. Surely, it is argued, we 'owe it to future generations' to make sure that they have as wide a choice as possible for responding to circumstances in fifty years' time that we cannot predict today. (In addition, of course, a certain level of nuclear skills will continue to be needed for a considerable period, for example to operate existing plants until they are closed down, to decommission those facilities and to manage or dispose of radioactive waste that has already arisen or will arise when plants are closed down, even if this should happen immediately.)

On another level, how could any major option really be 'closed' once the basic science and engineering have been developed? Closure and decommissioning of all today's civil nuclear facilities would not stop a revival at some point in the future – after all, civil nuclear energy grew from nothing directly after the Second World War to a position in which it provided almost one-fifth of the world's (much expanded) electricity use by the end of the twentieth century.

However, saying that an option should be kept open is very different from outlining what steps should be taken now to retain that option in various possible futures.

The inherent difficulty with any policy in areas of uncertainty is to determine what level of resources, if any, should be invested in developing a technology and associated infrastructure that may not ultimately be deployed. (The same argument applies to renewables, carbon dioxide sequestration and the more speculative ways of controlling energy demand.) Without investment the technology will not evolve to overcome problems or become more efficient, but of course investment diverts resources away from other possible uses. Investing in technologies, say for research and development, before there is a market demand is often likened to an insurance policy.

Moreover, the deployment of resources has a potentially detrimental effect beyond the lost opportunity to use the money elsewhere. The installation of demonstration nuclear facilities will create radioactive wastes and requirements for fuel cycle facilities and for transporting radioactive materials. Demonstration carbon dioxide sequestration projects could represent a potential local environmental hazard if large quantities of the gas were to leak from the store, for example. The local environmental impact of large-scale renewable projects could be considerable.

The opponents of nuclear power see it as a technology that has had its chance and has failed to deliver on its original promises, despite consuming enormous resources for some decades. No active steps, they conclude, should be taken to keep the option open, although research should continue to ensure that the legacy of waste and abandoned structures is managed as safely as possible.

To its supporters, nuclear power's potential as a contributor to plentiful, environmentally acceptable energy supplies is such that significant efforts should be made to overcome the barriers to its development. It is generally accepted that this will involve some level of government involvement, in creating frameworks for waste management, providing a stable regulatory environment and introducing some measure by which greenhouse gas emissions are penalized (for example a carbon tax). Supporters sometimes argue that some kind of government involvement in creating guaranteed markets for the power output of new nuclear stations (and renewables) may also be appropriate.

In addition, progress with basic problems in a number of areas may be necessary before further investment takes place, but the effort to overcome the problems may not be forthcoming unless demand is already evident. To take one example, demonstration plants of new designs of heavy capital plant are expensive. Private industry may not be prepared to fund these plants unless there is a reasonable prospect of subsequent orders. However, the investment necessary to support those orders may not be made unless the new designs are shown to be successful.

There may therefore be a role for government in creating sufficient incentive and support to ensure that plants are built to demonstrate whether or not new products will be attractive in prevailing market conditions. This issue is relevant not only in the nuclear field but also with reference to other energy technologies such as carbon dioxide sequestration, geothermal energy or large-scale bioethanol production.

Incentives and barriers to nuclear development

It should be noted that nuclear technology is not limited to the production of electricity. Reactors are needed to produce radioactive materials for use in medicine and a range of industrial techniques, such as non-destructive testing of aircraft components. The BN-350 reactor in Kazakhstan was used to desalinate seawater as well as to produce electricity. Nuclear reactors could be used to manufacture hydrogen by electrolysis or, perhaps, by direct pyrolysis of water. The use of nuclear reactors in propulsion is well established (submarines, icebreakers and even railway engines). Heat from nuclear reactors could be used in obtaining useful products from unconventional oil sources, such as tar sands or oil shale. Some of these applications could make nuclear power look more attractive if reactors were designed for dual-purpose use. However, the focus of this book is on the use of nuclear power to generate electricity.

Incentives for nuclear development

A number of factors outside the control of the industry might make nuclear power look more attractive than would otherwise be the case. They include:

- continuing very rapid growth in world energy demand;
- centralized, long-term approaches to power production systems;
- threats to security of fossil fuel supplies, perhaps because of growing international tension;

- growing concern about the environmental effects of fossil fuels, as climate change proves to be as serious as is currently projected by the Intergovernmental Panel on Climate Change (IPCC) and others;
- failure of renewable technologies to develop as rapidly or economically as is presently projected by their enthusiasts;
- failure to develop large-scale methods of storing electricity;
- retention or expansion of centralized power grid systems, with concomitant requirements for baseload power production.

In a world in which energy demand was growing rapidly, in which there were concerns about the availability of fossil fuels and their environmental consequences, in which electricity markets were centrally controlled rather than under competitive pressures and in which renewables appeared either impractical or uneconomic – a 'desperate world' – the 'need' for nuclear power might be so great that new nuclear stations would be built even if a number of current problems had not been solved. People might be prepared to accept a higher degree of uncertainty over waste management policy or higher generation costs, for example. To put it another way, the current state of development of nuclear technology would be sufficient for the option still to be open in 'desperate' futures.

Something approximating this world exists in some countries today, either because of rapid growth in electricity demand or because of profound fears about the security of energy imports. Of the five new reactors on which construction commenced in the year 2000, three were in large less industrialized countries (two in India, one in China) and the other two were in Japan. By the end of 2000, of the 33 reactors under construction globally, eight were in China, two in India and one in Argentina. Most of the rest were in the former Communist countries of central and eastern Europe and in the industrialized and industrializing countries of the Pacific Rim (three in Japan, four in South Korea).[3]

In contrast would be a 'complacent' world in which energy demand was not growing so rapidly, and eventually might even fall. In this world, fossil fuel use would be more acceptable, either because of falling concerns about climate change or because efficient and cost-effective ways of sequestering carbon dioxide emissions had been developed. Renewables would prove to be economic ways of generating large amounts of power, and localized embedded generation and competition in power markets would be the norm. Nuclear

---

[3] IAEA (2002).

power might be attractive only if the barriers listed below had largely been overcome – for example, if a clear waste management route had been identified and implemented and if costs were much lower.

In fact, it is unlikely that the whole world will be 'desperate' or 'complacent' at the same time. It is of course entirely possible that from one country or region to another, the situation might vary between the extremes, as is the case today. For example, some countries will have more energy options than others. The much higher growth rate of energy demand, and especially electricity demand, in some large less industrialized countries might make nuclear power seem more attractive, given the difficulty of installing sufficient capacity using alternatives, than it would be in those countries where power demand was only edging up and the requirement was mainly for replacement capacity as older plants were retired. Further, perceptions of global or national energy situations can change, sometimes quite rapidly. Some commentators argue that between 2000 and 2001, California went from complacency to desperation and back in less than a year.

Barriers to nuclear development

The issues described – overall energy demand, environmental issues associated with fossil fuels and the attractiveness of the alternative options – are largely beyond the control of the nuclear industry. However, a number of potential barriers to nuclear development are within its control, and/or within the direct control of governments, and it is here that the answer to the question 'What does keeping the nuclear option open actually mean?' is likely to be found. The following factors are relevant to overcoming those barriers.

Economic factors. To many commentators who are not instinctively opposed to nuclear power, the economics of nuclear stations of the type that have been built over the past two or three decades are the most important stumbling block to a nuclear revival in many countries. In the largely centralized markets to be found, for example, in some developing countries, large power reactors may well still be appropriate, but developing these reactors with much lower capital costs than has been the case in recent years would clearly make nuclear power relatively more attractive. In competitive power markets, smaller reactors, which are more suited to a system with more embedded capacity and which have shorter construction times, may be required as well. In addition, the regulatory regime, which has proved to be a source of much delay and uncertainty, especially in developed countries, may need to be stabilized.

Further, there is a big debate over whether the least industrialized countries would be better advised to develop power systems based on localized and embedded generation rather than on the large-scale grids typical of the industrialized world today.[4]

The fuel cycle. The decline or modest expansion of nuclear power could be fuelled for many decades on the conventional uranium reserves that have already been identified. However, a major expansion of nuclear generation would in due course require more uranium supplies, whether through better use of known reserves, fresh discoveries of conventional reserves or the development of unconventional reserves such as extraction from sea water. Alternatively, progress may be required in alternative fuel cycles, perhaps involving the use of plutonium, in fast reactors, thermal reactors (high-temperature gas reactors or mixed oxide-fuelled light water reactors) or accelerator-driven sub-critical reactors. Some commentators have suggested, however, that public acceptability of new reactor programmes would be more easily won if the more controversial aspects of current nuclear practice, such as reprocessing, were abandoned.

Waste. The demonstration of a practical and acceptable approach to waste management is important, both in terms of public acceptance and because in some places storage space for spent fuel is running short. A number of other issues contribute to this, for instance the debate between deep disposal and on- or near-surface storage, the role of reprocessing and the possibility of more speculative approaches such as partition and transmutation.

Safety. Studies[5] suggest that the safety record of nuclear reactors has been impressive. However, a major expansion of nuclear power could result in an increased number of incidents, all else being equal. Failsafe approaches to power production, to the extent possible, and a continuation of attempts to ensure the exchange of expertise and technology among user economies may help to reduce the risk of accidents.

Public perceptions. The state of public opinion about nuclear power varies among regions and countries. In general it seems to be more favourable in some larger less industrialized countries than in some OECD countries. Nonetheless,

---

[4] G-8 (2001).
[5] For example, IAEA (1991).

ensuring the nuclear industry is in proper touch with social trends and requirements, building trust and developing more democratic decision-making
procedures are likely to be of general value, and may be essential in at least
some countries. A particular challenge is that of finding ways of allowing potentially affected communities to take part in formulating policy at an early
stage.

Skills. A sufficient stream of suitably qualified personnel for reactor design,
operation, fuel cycle requirements, regulatory bodies etc. is a prerequisite for
continued nuclear development. Although the high social status accorded to
nuclear power in many developing countries (as was the case in its early days
in the developed world) has ensured a flow of personnel, in many developed
countries it is proving difficult to persuade undergraduates that nuclear power
is an attractive career. As a result many university courses in nuclear engineering have closed, and regulatory bodies are concerned about their preparedness
to license and monitor new reactor designs. This issue is of importance even if
nuclear power is seen to be in general decline since a number of tasks, such as
operating and maintaining stations in their final years, decommissioning redundant facilities and managing radioactive waste, will require skilled personnel for
many decades.

R&D. Considerable resources might be necessary to discover if new approaches
to nuclear technology could be developed to address current questions. However, it is not clear who should provide the funds, or how funding should be
apportioned between different potential new energy options. This issue em-
braces all of the possible options for tackling energy demand and environmental
challenges over the next years and decades.

Weapons proliferation. A robust response to risks of nuclear terrorism and the
safeguarding of fissile materials will continue to be required. A growing
world nuclear industry might be expected to be a greater target, and the recognized danger of terrorists obtaining radioactive materials seems to have grown
since the events of 11 September 2001. The management of spent fuel and
separated plutonium prior to disposal may be an area of particular concern.

Listing the above requirements does not of course presuppose that progress
in any or all of them is actually possible. It may be that nuclear power cannot
develop in a way so as to overcome all of its difficulties and that the nuclear
option can be kept open only for the more extreme scenarios mentioned earlier.
The same point can be made about renewables and sequestration of carbon

dioxide. However, the more progress is made in developing and improving the four options (including energy demand reduction), the greater the likelihood that the long-term challenges of global energy demand and reducing greenhouse gas emissions will be met.

The key issues

This book provides a more detailed consideration of the main issues identified in Phase One of the project. The next six chapters cover:

* public perceptions and decision-making;
* the relative economics of nuclear power;
* waste management, decommissioning and proliferation;
* nuclear safety;
* nuclear research, development and commercialization;
* nuclear power and the Kyoto mechanisms.

The themes that emerge from these chapters form the basis of the final chapter, which offers proposals for action.

References

BP Amoco (2001), *Statistical Review of World Energy 2001*. London: BP.

Depledge, J. (2002), *Cimate Change in Focus: The IPCC Third Assessment Report*. London: RIIA.

G-8 (2001), *Final Report*. Renewable Energy Task Force.

IAEA (1991), *Senior Expert Symposium on Electricity and the Environment*, sponsored by WHO, UNEP, WMO, CEC, IAEA, IIASA, Helsinki, 13–17 May; Key issue paper no. 2, 'Energy sources for technologies for electricity generation', SM-323/2, 30 April. Vienna: IAEA.

IAEA (2002), *PRIS* database, *http://www.iaea.org/programmes/a2/index.html*.

2 Public perceptions and decision-making in civil nuclear energy

Introduction

In some developed countries, nuclear power is unpopular. Until this unpopularity can be overcome, nuclear power will not flourish, even if the case for it on other grounds is strong. It has been particularly difficult in many countries to find new sites for nuclear facilities. In addition, some neighbouring countries have fears about nuclear installations. As cases in point, Austria is concerned about the Temelín plant in the Czech Republic and Ireland has objections about Sellafield in the United Kingdom.

In its early days, nuclear power was broadly accepted, often with enthusiasm, in many of those countries where it is now unpopular. But even then, there was a group of people, including some scientists who had worked in the field, which expressed deep unease about nuclear technology, both military and civil. The *Bulletin of Atomic Scientists* was first published in December 1945, and in an early edition Albert Einstein wrote: 'The unleashed power of the atom has changed everything save our modes of thinking, and thus we drift toward unparalleled catastrophe.'

By the late 1970s and early 1980s, those sceptical about both the technology and its practitioners were having a significant effect on public opinion, aided by a number of factors. The construction times and costs of many plants were far higher than projected, and the performance of many plants was very disappointing. The accidents at Three Mile Island and Chernobyl exacerbated growing mistrust of the nuclear industry and its often vocal supporters in governments. This mistrust had its origin, at least in part, in the arrogance and secretiveness of nuclear spokesmen in many countries. The suspicion that the industry and its supporters were able, for example, to put undue pressure on regulators further damaged their public credibility. Critics of the industry often had no apparent vested interest to do so, while the industry's responses to criticism came to be discounted – 'They would say that, wouldn't they?' The passion that has surrounded the nuclear debate in recent years is to a considerable degree a legacy of those times and attitudes.

At the same time, perceptions of the availability of alternatives to nuclear power were changing. When global fossil fuel supplies were under apparent threat (notably in the 1950s and again in the 10 years from 1973), nuclear programmes were instituted in many countries with relatively little objection, at least by today's standards. The discovery of vast reserves of gas as well as oil, coupled with their relatively low prices and the development of the highly efficient combined cycle gas turbine by the mid-1980s, reduced the apparent need for nuclear power in many countries. Without pressing economic or supply security arguments for nuclear power, it is perhaps natural that the problems with the technology should come to the fore. Public opinion towards nuclear power today seems to be more positive in those countries with limited indigenous alternatives, such as Japan and France, and those with rapidly growing energy demand, such as China and India. It cannot be assumed, however, that a strengthening of the positive case for nuclear power, perhaps because of growing concerns about climate change or projections of shortages of hydrocarbon reserves, would be sufficient to bring nuclear power back into public favour in other countries.

The nuclear debate is between two groups, the advocates and the opponents, who, despite their opposed views, have many attitudes in common (see Table 2.1).

An unwillingness, or perhaps even an inability, to engage in meaningful discussion with stakeholders also characterizes the extremes of the debate. There have been many examples of nuclear advocates claiming, often with apparent frustration, that 'the public' simply does not understand how beneficial nuclear technology is, while making little apparent attempt fully to understand its concerns. A similar attitude is to be found among the industry's adversaries. In reply to a request for information for this project, the authors received an e-mail saying: 'As you might imagine, our [Greenpeace UK's] starting point is that nuclear power doesn't have a future. There is therefore little point in talking about it.'

Most people holding a view on nuclear technology fall between these extremes, and are more prepared to accept the strengths and weaknesses of both sides' views. Many of those who are not firmly committed to one side or other of the nuclear debate show a willingness to move their position, for example as new information becomes available.

There is evidence that public relations departments in the nuclear industry are learning from the past and are moving from the 'one-way' model of communication that was prevalent in the 1950s towards a more balanced two-way approach; this suggests the industry's willingness to compromise where compromise is a feasible way forward. Similarly, regulators and decision-makers are now well aware of the range of views that face them. However, the level of emotion employed by the diehards on both sides of the debate is unlikely to

Table 2.1 Attitudes towards nuclear power: advocates and proponents

The advocates	The opponents
Belief that major elements of the future are predictable; certainty about general projections for various energy sources. (For example, renewables demonstrably do not have the practical potential to be other than relatively minor players in world energy supply.)	Belief that major elements of the future are predictable; certainty about general projections for various energy sources. (For example, renewables demonstrably have the practical potential to predominate in world energy supply.)
Certainty about the future major and important role of nuclear power and about issues such as nuclear waste (not a difficult technical problem).	Certainty about the future role of nuclear power (no role at all) and about issues such as nuclear waste (a technically insoluble problem).
Arrogance born of belief in infallibility of own analysis.	Arrogance born of belief in infallibility of own analysis.
Belief that the public is irrationally frightened of nuclear power. If only people could be properly educated, they would become more pro-nuclear and support the nuclear industry.	Belief that the public is irrationally complacent about nuclear power. If only people could be properly educated, they would become more anti-nuclear and support anti-nuclear campaigns.
Characterization of opponents as either foolish or ill-intentioned.	Characterization of opponents as either foolish or ill-intentioned.
Belief that government is not to be trusted to take wise decisions as it is too much influenced by the anti-nuclear media and pressure groups.	Belief that government is not to be trusted to take wise decisions as it is too much influenced by the nuclear industry and its supporters.

help in reaching careful and considered decisions about the future of nuclear power. Ultimately decisions must be taken, and those decisions will not satisfy everyone.

Public opinion

Considerable caution must be exercised when interpreting 'public opinion'. The very concept 'public' is of limited value in a modern pluralistic society. The population is better viewed as an interlocking pattern of smaller 'publics'. Moreover, individuals may move from one public into another if, for example, proposals are made to construct a major nuclear project near their homes.

A useful distinction can be made between 'opinions', 'attitudes' and 'values'. Worcester[1] offers an oceanographic analogy: opinions – ripples on the surface; attitudes – the tides; and values – the deep currents. Thus the results of opinion polls are notoriously dependent on the particular question asked. Reflecting the opinions they gauge, polls too can be very volatile.

Whether a particular person or group of people tends to be pro- or anti-nuclear at a particular moment depends on a number of factors, including:

- perceptions of the 'need' for the technology. As already noted, nuclear power tends to be more popular in countries with serious concerns about energy security than in countries with a wide range of energy options and a relatively modest growth in energy demand;
- perceptions of risk. Nuclear power tends to be less popular after an accident, especially if it occurred locally or had local consequences, or after other major events that may be relevant to the safety of nuclear installations, such as the terrorist attacks of 11 September 2001. But people who are more familiar with the technology, perhaps through living near a plant for some years, tend to be less worried than those who are not;
- social, political and psychological considerations. Political parties in a country can hold radically different views on nuclear technology, for instance the CDU and the SPD in Germany. Individuals who are attracted to large economic or technological projects tend to be more pro-nuclear, or at least to be less impressed with the arguments of anti-nuclear pressure groups, than people who are suspicious of globalized markets and 'capitalism' in general. People whose jobs depend on the local nuclear facility tend to be more pro-nuclear than those whose jobs do not.

It should also be kept in mind that there is always a range of 'opinions' among people from the same country and apparently very similar backgrounds.

A number of specific explanations have been suggested for the apparent special unease felt about nuclear power in many countries. They include:

- its links to the military, both real (the development of shared facilities) and perceptual;
- secrecy, coupled sometimes with an apparent unwillingness to give 'straight answers' (in part, perhaps, because of links to military nuclear operations in some countries, in part because of commercial issues);

---

[1] Worcester (2001).

- the historical arrogance of many in the industry, who dismiss opposition, however well-founded or sincerely held, as 'irrational';
- the apparent vested interest of many nuclear advocates, in contrast with the apparent altruism of opponents who, for example, are often not funded to take part in public inquiries;
- the perceived potential for large and uncontainable accidents and other environmental and health effects, notably those associated with radioactive waste;
- the overselling of nuclear technology, especially in its early days and in particular with regard to economics, leading to a degree of disillusionment and distrust;
- a general disillusionment with science and technology and with the 'experts know best' attitude prevalent in the years immediately after the Second World War;
- the wider decline of 'deference' towards 'authority' (including politicians and regulatory bodies).

Many of these factors are relevant in other public policy debates, notably the controversies over genetically modified organisms, road-building and global trade. The unease about such technologies is considered in more depth later.

Perceptions of negative public opinion, whether justified or not, can be extremely expensive for investors in nuclear power. Opposition to the construction or operation of nuclear facilities can increase the costs of nuclear-generated electricity in a number of ways. There may be delays during construction or in achieving an initial operating licence or there could be interruptions in operation. Extra physical or operational security measures might be demanded, say in response to a potential terrorist attack, even if there is no direct evidence of a threat. Implementing these measures may be especially costly if they involve 'backfitting' an existing design. The costs of site selection, evaluation and the licensing process itself can increase. The costs of transporting nuclear materials can escalate, because of increased requirements for security against protest or the need to find new routes. The economic risk associated with uncertainty leads to demands for higher rates of return on investment, an especially serious issue for highly capital-intensive technologies such as nuclear power (see Chapter 3, on the relative economics of nuclear power).

In the most extreme cases, fear of public reaction can result in the refusal of an operating licence to a fully completed plant or in a government preventing nuclear construction or closing down existing facilities before the end of their technical lifetime. Thus, since 1978 a total of 21 nuclear power plants, with a

combined capacity of some 14 GW, and one mixed oxide (MOx) fuel production plant have been closed or halted in advanced stages of construction for
non-economic reasons in six OECD countries (Austria, Germany, Italy, Spain,
Sweden and the United States), some as a direct result of referenda. Most of
these closures were carried out in the years after the Chernobyl accident in
1986, although the 720-MW Tullnerfeld reactor in Austria was refused an
operating licence on completion after a referendum in 1978. Italy no longer
operates nuclear power reactors, having closed three operating plants after a
referendum in 1987. Germany, the Netherlands and Sweden have adopted formal phase-out policies by law, Switzerland adopted a 10-year moratorium on
new construction in 1990 and Belgium has taken a policy decision to phase out
nuclear power. A number of countries that do not have operating nuclear
power plants, such as Australia, Austria, Denmark, Greece, Ireland, Norway
and Poland, have put in place legal or policy obstacles to nuclear power.[2]
Whether or not the public in these countries really is deeply suspicious of
nuclear technology and whether or not its fears are justified, perceptions about
the public mood have had, and will presumably continue to have, profound
implications for nuclear power's development or failure to develop.

The complexities of 'public opinion'
It does seem that when asked whether new nuclear power stations should be
built, people in most developed countries tend to say 'no'. However, the same
applies to a number of other industries that seem to continue relatively unfet-
tered. In the United Kingdom, for example, the chemical industry regularly
appears less popular than the nuclear power industry in opinion polls (when
people are asked which industries they believe cause environmental damage).
However, with certain exceptions such as production of PVC, most chemical
companies have not seen their business severely curtailed to the extent of the
nuclear rejection listed earlier.

On the other hand, people seem to have little difficulty in responding posi-
tively to the public image of a particular firm while disliking or mistrusting
the industrial sector in which that firm operates. The phenomenon of holding
apparently contradictory attitudes towards something is common in the field
of public perception. A number of technologies, or indeed the concept of technology itself, can be both exciting and frightening. The fact, mentioned on
page 17, that the wording of a question can be crucial in determining the

---

[2] IEA (2001).

response is well known to opinion pollsters. Questions about whether nuclear power is desirable 'as fossil fuels run out and concerns over global warming increase' are far more likely to receive positive responses than questions about whether nuclear power is desirable 'given the risks of a major accident like Chernobyl and the insoluble problem of nuclear waste'.

Also, public attitudes, or at least public responses to questionnaires, can change rapidly. This was demonstrated in California at the time of the power crisis of 2000/2001. As the power cuts bit, support for new nuclear stations rose significantly, to fall away just as rapidly when the power shortages eased.

'Public opinion' or 'public confidence' would thus seem to be heavily influenced by the context in which questions are put. The degree to which support or opposition is expressed, however, is only one aspect of an opinion. The strength with which the opinion is held and the likelihood that the topic will be mentioned without specific prompting are also important.

In most developed countries there is little evidence of a high level of overt public concern among people not directly affected by nuclear projects. When people are asked, unprompted, about their concerns about the environment, nuclear power comes some way behind urban pollution from motor vehicles and climate change. Public concerns that are widespread but 'back of mind' are especially difficult to evaluate. (Some recent attempts to address these difficulties are described in Appendix 2.1.)

Perceptions of public perceptions

When it comes to considering the effect that public opinion has on decision-making, a further complication occurs. Decision-makers, naturally, will base their decisions in part on their perception of public opinion, in other words on a perception of a perception. There is some evidence that these second-order perceptions may also be subject to error.

As indicated in Table 2.2, opinion polling carried out by MORI[3] in the United Kingdom suggests an interesting pattern of perceptions about public perceptions of nuclear power among opinion-formers and decision-makers. (A similar pattern has been observed in the United States.) These data imply that at least in some countries, decision-makers' perception of public opinion may not be accurate and therefore that decisions, to the extent they are made with regard to public opinion, may be skewed by assumptions that are not true. The possible reasons for this include the attitude of certain elements of the

---

[3] BNFL (1999).

Table 2.2 Attitudes towards nuclear power among decision-makers

	Favourable towards nuclear energy industry (%)	Unfavourable towards nuclear energy industry (%)	Neither favourable nor unfavourable/ don't know (%)
Public opinion	28	25	47
All MPs	43	44	13
MPs' perception of national public opinion	2	84	14

popular media and the greater effectiveness of anti-nuclear pressure groups in organizing letter-writing and other publicity campaigns.

There are other paradoxes too. Opinion polls in Sweden, for example, show a large majority in favour of the continued operation of existing nuclear power stations,[4] a view also taken by most of Swedish industry. Nonetheless the governing coalition forced the closure of one nuclear plant in 1999, and another is to follow (at least in principle). The internal politics of governing coalitions may also be a factor in determining policy in countries such as Germany.

The above observations emphasize the need to explore and evaluate public views at an early stage in the decision-making process. This is the aim of some of the innovative techniques discussed later.

Public attitudes to particular projects

Attitudes towards nuclear technology, the values on which they are based and how both are perceived by decision-makers are very important. For example, companies considering investing in new nuclear power stations may be more willing to do so if they consider 'the public' to be broadly supportive. But if the public is clearly sceptical about nuclear technology, fears may arise about changes in government policy that would work against the industry – a case in point would be the possibility of a long delay between the completion of a plant and the granting of an operating licence or the imposition of stricter emission limits. Even if this did not deter investment entirely, it might well lead to higher rates of return being demanded in compensation for the higher risks involved.

---

[4] Analysis Group (2000).

However, even if in a particular country there were little evidence of front-of-mind public opposition to nuclear power in principle, proposals to build a particular facility at a particular site might provoke a strong backlash from the local population and others. Widespread public action against waste transportation in Germany and shipments of plutonium to Japan indicate that many people, who might regard themselves as unconcerned about nuclear power when not directly involved, can change their view and behaviour significantly when confronted by a concrete proposal, especially if the object of that proposal is nearby. Probably at least some of the opposition to construction proposals is prompted by a feeling that local people have not been properly involved in the decision-making process or that the risks and benefits associated with the project are not being fairly distributed. Various ways of addressing these matters are discussed later. However, even if these feelings can be addressed, there is no guarantee that a broad consensus can be reached with respect to any particular scheme.

A brief history of early nuclear perceptions

For some individuals, fears about nuclear power are more acute than fears about other risks that, based on the past record, are more dangerous. Prime examples are the use of coal for electricity or petrol for transport, each of which has been associated with many deaths through the health effects of emissions. Further, the debate about nuclear power has an emotional tone that is unusual even within the controversial and uncertain field of energy, an observation which has prompted this project. The fascination of some elements of the mass media with matters connected with radiation may be part of the explanation, but it is unlikely to explain the observation fully.

Some commentators, notably Weart,[5] have observed that radiation and nuclear energy have a number of properties that connect with deep-seated, apparently universal human images and myths, both positive and negative in their emotional tone. Elements of this 'pre-fission nuclear imagery' include:

- 'Eden' myths: tales of a golden age before the fall of man was brought about by meddling with forbidden knowledge – the 'small boy playing with fire';
- images of the destruction of mankind, often by fire. The concept of 'atomic bombs', for example, was introduced in H. G. Wells's novel of 1913, *The World Set Free*;

---

[5] Weart (1988).

- the 'taming of nature' and the unleashing of massive forces for good or evil;
- the alchemists' dream. The transmutation of elements, which occurs during radioactive processes, has held a fascination for many centuries;
- the idea, found in many civilizations, of invisible rays, undetectable by unaided human senses, that could bring life or death and could blight future generations.

Before the Second World War, debate about the use and effects of radioactive materials was not widespread. Before 1945 there did not seem to be any particular fear of radiation. It had benefits, especially in medical treatments, and risks, and many people seemed happy to believe that the former outweighed the latter. High levels of trust in the scientists involved might also have been a factor.

The atomic weapons dropped on Japan at the end of the Second World War had a profound effect in many ways. They revealed to the public that atomic energy was a reality, not merely an ancient dream. The image of world destruction was all too believable in the face of photographs and reports from Hiroshima and Nagasaki. Post-war programmes such as civil defence seemed to cause even greater anxiety among the population. Concerns both about the idea of nuclear technology and about the practices of some of those in charge of it were raised by some of the very scientists who had been involved in its development, perhaps most notably Albert Einstein, Robert Oppenheimer and Joseph Rotblat. The danger became clear that fear in the population could result in the rejection of nuclear technology, notably weapons, that the US government, and increasingly other countries, had begun to regard as essential. Partly to reduce such pressures, scientists in many countries, particularly the United States, spoke of their determination to turn the dreadful power of the atomic weapon into a positive force by using it as an alternative way of making electricity and providing propulsion.

The American programme 'Atoms for Peace' was launched by President Eisenhower in 1953. Under its auspices, the US Atomic Energy Commission staged exhibitions displaying the benefits of atomic energy throughout the 1950s in places such as Karachi, Tokyo, Cairo, Sao Paulo and Teheran. 'Atoms for Peace' was officially promoted as a step in controlling nuclear weapons proliferation. Important help in developing nuclear technology was offered to any country that would allow international inspection of its facilities in order to verify that materials were not being diverted for military uses. Another purpose of this programme, however, was undoubtedly to reduce the level of potential public opposition to the continuance of the US nuclear weapons programme in the face of the Soviet nuclear threat. The American people as whole were to be made aware of the benefits of nuclear technology. By 1955, a secret report to the president confirmed that 'Atoms for Peace' had 'detracted popular attention away from the image of a United States bent on nuclear holocaust' and diverted the public eye to 'technological progress and international cooperation'.[6] Another motivation, in the early years of the Cold War, was the need to persuade countries to choose US rather than Soviet nuclear technology.

There was strong advocacy for nuclear power in other countries as well. In Europe and Japan the most easily mined coal was gone, the best hydropower sites were in use and almost all oil was imported. The energy shortages of the war were followed by coal crises in the post-war years. In the extremely severe winter of 1947 much of Europe faced shortages of fuel to heat homes and electricity to light them, while Japanese cities faced brownouts well into the 1950s.

On the other hand, owing to the highly complex nature of the subject even many politicians felt unable to engage meaningfully in the details of the debate. Responsibility not only for implementing nuclear policy but also for setting it was to a considerable extent vested in bodies representing nuclear experts. Inevitably, in the view of some commentators, a technocratic mode of decision-making became dominant, to the detriment of dialogue with and control by normal democratic institutions. The secrecy associated with military uses of nuclear materials exacerbated this tendency. It is perhaps unsurprising that in these circumstances the spokesmen for the nuclear industry became increasingly arrogant.

The ground was fertile, therefore, for nuclear power to be oversold as the new miracle fuel. The rate of publication of popular books and articles on nuclear power trebled after Eisenhower's 'Atoms for Peace' speech. Prior to an international conference in Geneva in 1955, most nuclear scientists had warned that as an economic venture, nuclear power was still decades away. But the 3,000 scientists and their followers attending the conference heard from representatives of several countries that nuclear power was close to being commercially profitable. The American people were told, famously, that nuclear power would become 'too cheap to meter', an incomprehensible but highly influential statement. The enthusiasm of many more countries, such as West Germany, for the new technology was fired, while the Suez crisis of 1956 added urgency to the perception that a new source of energy was required. By the end of 1957 the United States had signed bilateral agreements on nuclear

---

[6] Possony (1955).

technology with 49 countries, and American firms had exported 23 small research reactors. The Soviet Union had also invested enormous resources into developing nuclear technology for export within the Communist bloc, both as a Cold War challenge to the United States and through a similar belief in a nuclear utopia.

Often, however, enthusiasm for new technologies and other human endeavours is exaggerated, and disillusionment sets in. The Atomic Scientists of Chicago (named after the site of the first controlled fission experiment in 1942) was formed in 1945 in order to sound warnings against believing all of the official pro-nuclear pronouncements. Over the next three decades further groups were founded that argued this case. Perhaps the most notable were the Campaign for Nuclear Disarmament (1957) and Greenpeace and Friends of the Earth (1971).

Attitudes and feelings

It has been assumed, at least by physical scientists and at least until relatively recently, that our attitudes to issues such as the use of nuclear power derive from our interpretation of the facts presented to us. Thus the nuclear industry considered that public fears about, say, nuclear waste, which arose because of ignorance or irrationality, could be countered by 'education'. The 1980s offered many examples of full-page advertisements about nuclear waste, reactor safety, the health effects of radiation and the like, all designed to increase the public's knowledge, and thus acceptance, of nuclear matters.

The level of public knowledge about nuclear power is low. A public survey carried out in the United Kingdom in the mid-1980s[7] found that the best-known fact about nuclear power was that uranium was its fuel and that this was known by 31 per cent of respondents. Other surveys have shown that some 50 per cent of people believe nuclear power to be responsible for acid rain and that only 13 per cent of people are aware that radiation can come from both natural and man-made sources. Nonetheless, as has become increasingly clear, the analysis that opposition to nuclear power is caused by low levels of knowledge is at best simplistic. One need only reflect that many of the 'professional' opponents of nuclear power are extremely well versed in the issues.

The gap between the 'rationality' of nuclear scientists and the 'rationality' of the public is shown in the debate about nuclear waste. Most people in the nuclear industry judge that radioactive waste management and disposal offer

---

[7] Lee et al. (1983).

no insuperable or interesting technical challenges. It has even been said that radioactive waste is the only long-lived waste stream for which a solution exists. Intermediate-level waste (ILW) is described as a material that is about as dangerous as paint stripper or petrol. It could certainly cause problems if not treated carefully, but with proper safeguards it can be handled perfectly safely. But while scientists are explaining why ILW is no more dangerous than many other types of waste, and less dangerous than some, at the same time they explain that this material requires burial 800 metres underground in an area with very little groundwater flow and without deposits of valuable minerals in case people dig up the waste by accident some centuries in the future.

These messages are clearly contradictory. Burying waste deep underground is not a way of dealing with other industrial waste streams. The 'rational' assumption is therefore that nuclear waste must be far more dangerous than any other type of waste and, furthermore, that the nuclear scientists must be insulting the public's intelligence by pretending it is relatively benign.

The response of the nuclear industry has been to propose ever 'safer' solutions to radioactive waste management, on the assumption that people will be reassured when they see better steps being taken to contain the waste. In reality, however, the 'rational' response on the part of the non-expert would be to assume that radioactive waste was even more dangerous than had been admitted before, especially if regulators were producing progressively more stringent safety conditions. Fear of radioactive waste would be heightened, as would mistrust of the nuclear scientists.

Furthermore, assessments of the damage caused by low levels of radiation have changed in recent years. It is now believed that the survivors of Hiroshima and Nagasaki received lower radiation doses than was first thought. It has therefore been necessary to revise upwards calculations of the damage done by a particular dose of radiation. Again, the 'rational' response might well be to ask how many more times scientists will discover, or perhaps 'admit', that they have underestimated the problem before the 'truth' is told.

It seems, then, that the rationality of physical science and the rationality of everyday life can diverge quite radically. It seems likely that this is heavily influenced by our evolution. For example, there would have been little evolutionary purpose in being able to comprehend timescales longer than about 100 years (roughly the combined lifespan of ourselves and our children) or less than a modest fraction of a second or to appreciate distances greater than a few kilometres or smaller than a millimetre. Yet scientists work with such concepts all the time, although whether even they 'comprehend' them, as opposed to being able to manipulate representative symbols, is not clear.

This divergence between 'realities' is especially important when it comes to considering our reaction to low-probability, high-consequence risks. (For the sake of the following argument, let us ignore the legitimate observation that the calculations of riskiness are of their nature less reliable for very low-consequence events and assume that the following risks are well established.) If a particular activity has a one-in-ten chance of being associated with an accident in which 10 people are killed in a single day, it is relatively easy to understand the safety implications of that activity. However, if a risk has a one-in-a-thousand chance of killing 500 people in a single day – and would therefore cause only half the number of deaths over a long period of time – our response may still be one of greater unease. (The consequences of a major rail accident, for example, tend to be more significant than of a series of road traffic accidents, even though rail travel is demonstrably far 'safer'.) 'One-in-a-thousand' is at the very edge of our conceptual ability to appreciate it. If an activity creates a one-in-a-million chance of killing 100,000 people in a day or a one-in-ten-million chance of killing 500,000 people, concerns are likely to be greater still. As 'one-in-a-million' and 'one-in-ten-million' are literally inconceivable figures, and thus are psychologically indistinguishable, the focus will inevitably be on the potential consequences.

Everyday rationality places far more value on human factors. Churchill once said that when deciding what weight to put on a statement, he would first consider who was saying it, then how they were saying it and finally what they were saying. In everyday life, most of us seem consistently to make up our mind about many matters as a result of whether we like and, especially, trust the messenger than of critically examining the message.

The ideology of science, on the other hand, is different. When a piece of research has been published after suitable peer review, it can be assumed to be the result of an objective examination of the facts. It is relatively recently that this ideology has been seriously challenged, in view of the observation that scientists too have their vested and psychological interests. There seem to be plenty of cases – the early research carried out into the health effects of tobacco is a case in point – where scepticism about the objectivity of some of the scientists involved, or at least the extent to which they were prepared to make public statements detrimental to their paymasters, has proved to be well founded.

Apart from scepticism about the motivation of any particular scientist, there is a gap between the way scientific culture and political-media culture each views science. Scientific 'knowledge' can only ever be provisional. However well a phenomenon may be examined, the possibility that new observations will become available that cause a revision of existing hypotheses can never be dismissed. A major example was the replacement of Newtonian mechanics with quantum theory in the early years of the twentieth century. Less dramatic cases are plentiful. Especially when it comes to considering new and/or very rare phenomena, one would expect that early conclusions may well be over-turned as data accumulate.

Politics, and in a different way the media, tend to be more 'black and white' in the ways they operate. Decision-makers sometimes require simple answers to essentially complex questions, of which perhaps 'Is it safe?' is the most common. Genuine scientists tend never to express their responses in words such as 'irrefutable' or 'unequivocal', but the political establishment often responds as if these terms have been used. There is suspicion in some quarters that scientists themselves have not always worked hard enough to dispel the aura of infallibility, perhaps because of the personal kudos it can bring.

'Science', then, has often been perceived by many in the public, the media and politics as a source of 'right' or 'wrong', but always 'absolute', answers. As new discoveries and complex scientific debates result in a change in accepted scientific understanding from time to time, there has been disillusionment with science itself. Yet it is difficult to believe that decision-making in the twentieth century would have been better had scientists been refused a contribution to that process. To improve the understanding of what science can and cannot offer the decision-making process would be of great help in many complex areas.

Factors affecting our perceptions of risk

Society's attitudes to risk sometimes appear to be paradoxical, at least on first view. Although risk is clearly regarded as undesirable and to be avoided in most cases, there are various instances, such as 'extreme sports', of risk being valued for pleasure; and individuals who take major and apparently unneces-sary risks are often admired.

The term 'risk' has a fairly clear technical meaning to the safety profes-sional; it is the amount of harm (measured in deaths, or sometimes injuries) associated with a given quantity of the activity in question (say 'per 1,000 participants per year' or 'per individual per 1,000 hours'). But the public does not seem to interpret 'risk' in the same way as do risk professionals.

Oughton[8] asked individuals to respond to two identical lists of activities and rank them in response to two questions: 'Which of the following do you con-sider to be the most risky?' and 'Which of the following do you associate with

---

[8]   Oughton (2001).

the highest probability of premature death?'. (See Table 2.3.) Different activities were mentioned most frequently in the two cases. (Bold text indicates hazards with a statistically significant likelihood of being mentioned more frequently than the others.)

These results suggest that when we refer to an activity as 'risky', we refer to more than simply its perceived potential for causing death. In part the explanation may lie with the potential for causing non-fatal injury, but this is unlikely to be the whole story. It may be that some aspect of particular risks leads people to feel more uncomfortable about them than others, given the same likelihood of causing harm. Alternatively, there may be something about the social context of certain risks – for example, a particular mistrust of the individuals associated with the activity – that may lead people to regard the activity as 'risky', even when they are aware that it is unlikely to be associated with direct harm.

A second point concerns uncertainty, especially about new technologies or new circumstances relevant to established technologies. The aircraft industry has a very large database on the failure of aircraft designs; this allows the likelihood of accidents to be estimated with some confidence. But the precise likelihood of a major nuclear accident, like other very low-probability, very high-consequence events, cannot be calculated in this way. The estimates have to be constructed from calculations of the failure rate of individual components, and are inevitably less reliable as a consequence. (It can be very difficult, for example, to be certain how the various components may interact in an emergency or how operators may behave.)

Table 2.3 Attitudes towards potential risks

'Which of the following do you consider to be the most risky?'	*'Which of the following do you associate with the highest probability of premature death?'*
Nuclear power	**Smoking**
Genetically modified food	**Alcohol**
Food additives	**Car accidents**
Car accidents	**Air pollution**
Weapons	Bad diet/lack of exercise
Smoking	Weapons
Alcohol	Pesticides
Aircraft accidents	Aircraft accidents
Skiing	Nuclear power
Pesticides	Skiing
Radon	Food additives
Air pollution	Radon

Furthermore, the threat represented by a nuclear (or any other) facility involves not only problems with the internal working of the plant but also the possibility of external threats. Many kinds of attack are taken into account during the design of nuclear facilities, but it is impossible to be certain that all eventualities have been identified. The terrorist attacks on the United States in September 2001, for example, represented an apparently 'new', and unquantifiable, threat in the nuclear safety equation. There are parallel uncertainties concerning what the precise effects of a major release of radioactivity would be. They would almost certainly depend on factors such as the prevailing weather conditions at the time. Uncertainty inevitably leads to greater concern, even if the worst fears ultimately prove to be unrealistic.

Three kinds of factor might lead people to perceive risks of like magnitude differently from one another:

* factors associated with the particular nature of the risks themselves;
* those related to the social context in which risks are communicated;
* those associated with different types of personality.

The nature of different types of risk

Slovic et al.[9] investigated the factors that act between calculated risk, as determined by probabilistic safety assessment (PSA), and public perceptions of risk. (Probabilistic safety assessment, in effect, makes estimates of the risk associated with a particular activity by examining the dangers the activity has represented in the past. Results are often quoted in terms of 'deaths per mile' or 'injuries per hour' etc.) They asked various groups of people – the (US) League of Women Voters, college students, active club members and 'experts' (who used PSA) – to rank 30 potential risks in order of severity.

The correlation between the public perception of risk and the calculated risk was in general very good. However, in some cases there were considerable discrepancies. Thus, college students ranked swimming as the least serious of the risks (below food colourants and hunting), while in fact over 100 swimming deaths occur each year in the United States. By contrast, all groups put nuclear power near the top except the PSA experts, who placed it twentieth.

Slovic concluded that three factors come into play when we convert 'real' risk into 'perceived' risk. The first is familiarity. If a risk is an old, well-established one, familiar to the individual and easily detectable by unaided human senses, it

---

[9] Slovic et al. (1980).

will tend to be underestimated compared to a risk of the same actual magnitude that is new, unfamiliar and difficult to detect. The second is controllability. If a risk is run voluntarily and is easy to control, it will be perceived as less serious than one that is imposed on people and is difficult to control. (Note that this is not a matter of the *acceptability* of a self-imposed risk against one imposed on us but of the perception of the actual *severity* of the risk itself.) The third concerns the number of people affected by the risk. If a risk represents a small chance of damage to a large number of people (especially future generations), it will be perceived as more serious than a risk that has the same overall health effect but whose likely victims one can identify.

Car travel, responsible for several thousand deaths each year in a country such as the United Kingdom, lies near one extreme in all three factors. We are all familiar with car travel, having taken part in this activity without suffering a fatal accident. We choose to indulge in it, and we all know that we are better drivers than the majority and so are less likely to suffer an accident. Finally, although there are many deaths on the roads, it is always possible to identify the victims, and indeed relatively few such fatalities are reported in the mass media.

Nuclear power appears at the other extreme. It is a relatively new risk. Radiation is not detectable by unaided human senses, and in a number of countries most people usually see images of nuclear stations in negative contexts, the mass media generally preferring 'bad news' stories to good ones. It seems to be 'imposed' on local communities and on society at large; few people choose, or would choose, to have a nuclear power station built nearby in the sense that they choose to go for a drive. And although nuclear accidents are rare – there has been only one with demonstrable off-site health consequences caused by radioactive releases – the impression is that a single extreme event could affect large numbers of people, perhaps everyone on earth and for many generations, and it might even interfere with the genetic stuff of life itself. For example, many individuals in northern Europe were aware that fallout from Chernobyl fell on their homes, especially during rainfall, during the week after the accident.

As a result, nuclear power tends to cause more unease than motor transport, even among some people who suspect they are being 'irrational', in the same way that air travel tends to cause more anxiety than car travel, at least for many less frequent travellers.

In addition, perceptions of benefit are relevant to our perceptions of risk. In the case of car travel, the benefit, a convenient and private journey, is delivered immediately. This contrasts to, say, rail travel, where we have to get to and from the station and share the carriage. There is thus something of a psychological 'vested interest' in underestimating associated risks. The perceptual link between nuclear power stations and the electricity coming from a wall socket is somewhat esoteric by comparison, as are the claimed benefits in terms of, say, reducing climate change.

The social context of risk

The above analysis cannot explain fully how risk is perceived in everyday life. It is clear that other factors, such as the way in which 'information' is disseminated and the apparent motivation of those giving the information, are also significant.

It is undeniable that some of the technical issues in the nuclear field are so arcane as to be inaccessible to non-specialists. Furthermore, 'experts', people who clearly understand the field in great detail, disagree publicly about certain aspects of the technology. Thus, when people seek to make up their mind about the desirability or otherwise of nuclear energy, the credibility they accord to the source of information becomes a very important issue.

There is evidence that 'the public' has become much more sceptical about claims made by industry and government concerning complex technologies, a phenomenon often referred to as a 'decline of deference' (see Figure 2.1). People are more likely to believe statements by scientists working for environmental organizations than by scientists working in industry or government.

In its early days, the nuclear industry was, as noted earlier, heavily influenced by the views of technocrats. There was relatively little scrutiny by the political establishment, which often seemed to accept that it was simply not equipped to frame suitable questions or to understand the answers. 'Communication', when it was deemed appropriate at all, was a one-way process of informing the public of intentions.

The poor outcome of some of the industry's decisions, especially in commercial terms, coupled with its secrecy and intimate relationship with government, has left a legacy of mistrust. The result is that its pronouncements are treated with some scepticism by significant sectors of the population, and scepticism figures importantly in evaluations of the role of government as regulator. There is suspicion in some quarters about a cosy relationship between the industry and government which makes proper rigour in regulatory processes difficult to guarantee. In reality, it is much more difficult to identify large pronuclear factions within government in many countries today than it might have been 30 years ago, and, as noted earlier, a number of governments have explicit

Figure 2.1 Percentage of respondents often or always trusting institutions to tell the truth

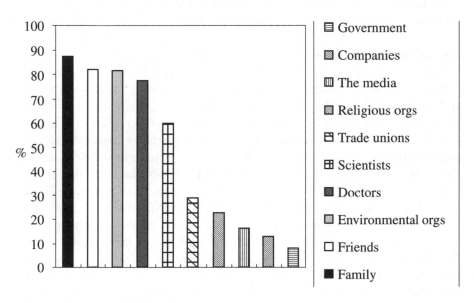

Source: Marris et al. (1996).

or implicit policies of nuclear phase-out. However, perceptions, even when based on evidence from some years ago, can often be difficult to overturn.

Communication will probably be most credible if its source is seen to be open, accountable, inclusive and equitable.

There is evidence that the nuclear industry has recognized the importance of becoming a genuine partner in decision-making with other stakeholders and with society at large. This implies the inclusion within the consultative process of representatives from a wider range of interests than was previously the case. Nonetheless, it may take some time before this change in attitude is recognized and accepted by elements of the public, the media and the political establishment.

'Personality' issues

The very fact that people from apparently similar backgrounds, subject to the same information and living within the same country, can have radically different views on an issue such as nuclear power suggests that personality factors may be important in forming attitudes. The responses to the Three Mile Island accident are instructive. The pro-nuclear community concluded that the incident confirmed the safety of nuclear power – 3,000 MW of heat energy went out of control without significant radioactive releases. The anti-nuclear groups saw it as an icon of the dangers of the technology.

Some systematic differences, notably that women are more anti-nuclear than men in many countries, may be difficult to explain in terms of the nature of the risk involved. The explanation may lie more in different attitudes to the health of the foetus or in the male-dominated public face of the industry. Those who oppose nuclear power have often been held to resemble the Romantics, who opposed the early events of the Industrial Revolution such as the growth of the railways and sought a simpler, more 'natural' life. Being anti-nuclear has often been equated with being 'anti-establishment'. Those who support nuclear power have often been portrayed as technocrats or capitalists, who believe fundamentally in the supremacy of man over nature.

There is some truth in these characterizations. However, in recent years there has also been an increasingly vocal group from the traditional Right which has criticized the nuclear power industry, perceiving it as an organ of a too-powerful state. Although this group tends not to be impressed by claims about the alleged detrimental environmental effects of nuclear power, its response is nonetheless often couched in forthright, even combative, terms. This group sometimes points to the international insurance regime, in which very large nuclear risks are covered not by the industry but by governments. The facts that nuclear risks are excluded from domestic insurance policies and that the industry is protected from very large claims by the terms of international conventions create the impression of an industry that is uniquely dangerous (though the very rapidly increasing rate of insurance claims resulting from instances of extreme weather, which may in turn be associated with climate change, could result in similar exclusions for storms, floods etc.).

In one study,[10] people were asked to classify themselves as 'pro-nuclear', 'anti-nuclear' or 'don't know'. All groups were then asked if they could name points in favour of and against nuclear power. All three groups tended to produce the same factors in their lists. 'Waste' and 'accidents' featured high among the drawbacks, while 'cheapness' and 'cleanness' were commonly cited advantages. However, the pro-nuclear group tended to mention each benefit more frequently than the anti-nuclear group, while the anti-nuclear group mentioned

---

[10] Lee et al. (1983).

each drawback more than the pro-nuclear group. (The 'don't knows' tended to mention fewer benefits or disadvantages than either the pro-nuclear group or the anti-nuclear group.)

These findings are by no means as obvious as they might first seem. Some of the benefits and disadvantages in question – that nuclear power does not contribute to acid rain or that there is no long-term waste management policy in the United Kingdom – are very close to being objective 'facts'. One might expect, then, that both pro-nuclear and anti-nuclear groups would mention these advantages and disadvantages equally if they were indeed making objective assessments of the facts and then making their mind up about their attitude towards the technology. It seems that, at least to some extent, the pro- or anti-nuclear attitude comes first, and then people select and filter facts to fit the attitude.

Those who described themselves as 'pro-nuclear' or 'anti-nuclear' broadly shared a number of characteristics. The 'pro' group, for example, tended to be older males who actively pursued a hobby of some sort, while the 'antis' tended to be female and younger. A small but measurable proportion of the 'pro' group could not mention a single benefit of nuclear power, while a similar proportion of the 'anti' group could not mention a single problem.

It would thus seem to follow that simply challenging, or even changing, people's understanding of the facts will not (always) change their underlying attitudes, to the extent that attitudes stem from more fundamental psychological factors. Studies tend to find relatively little difference in the level of factual knowledge among pro- and anti-nuclear members of the public. Providing technical information does not appear effective in building support. (Indeed, the emphasis on providing technical information may be counter-productive with respect to some sectors of the population. It may foster a sense that the industry has missed the point of people's real concerns and therefore that, literally and metaphorically, it does not speak the same language as the public.) Of course, those who oppose nuclear power would argue that raising the level of knowledge among the population would cause people to turn against nuclear technology.

Even a correlation between knowledge and a pro-nuclear position does not prove a simple causal relationship. People studying science in schools and universities tend to be more pro-nuclear than the population at large. However, it may be that they tend to 'like' nuclear (and other) technology and thus to wish to study science courses in order to find out more about it. Moreover, it appears that some who oppose nuclear technology will go to great lengths to find out about it, the better to be able to articulate an opposing view.

It is unlikely, even in principle, that it will ever prove possible to reconcile differences between the extremes of the pro-nuclear and anti-nuclear communities. In a Consensus Conference on radioactive waste held in the United Kingdom in 1999, for example, one witness argued against research into transmutation as a way of reducing the lifetime of long-lived radioactive wastes on the grounds that solving the waste problem would make nuclear power more attractive and should therefore be resisted.

The question 'What would it take to make you change your mind?' often elicits interesting answers. When asked what sort of solution to the waste management and disposal issue they would accept, anti-nuclear campaigners often answer that a prerequisite would be proof that the material will not leak into the environment while it is still measurably radioactive, i.e. for more than a hundred thousand years or perhaps ever. As discussed in Chapter 4, on radioactive waste, this response could mean one of two things. It could mean that the current level of scientific knowledge is not sufficiently great to allow for confidence that there will be no such leak (and thus that building a repository now would be unacceptable) but that research might in time give sufficient certainty about the future behaviour of a specific repository site. However, it could mean something quite different. Along the lines of Bohr's comment that 'prediction is difficult, especially about the future', it could mean that it will never be possible *in principle* to predict the behaviour of anything over such a long timescale. As it would never be possible to validate either the basic science or the detailed knowledge of a particular site, so would it never be possible to validate the site's suitability for the deep disposal of radioactive waste.

The corollary for the anti-nuclear community, of course, is that no more radioactive waste should be produced. Paradoxically, an exception is sometimes made for waste arising from non-power uses of radioactive materials such as nuclear medicine. These activities produce wastes similar in nature, although smaller in volume, to those arising from nuclear power. There is an apparent contrast between the 'absolutist' opposition to the generation of waste from power uses and the 'cost-benefit' analysis that seems to be applied to waste from nuclear medicine.

Sometimes the argument that waste disposal can never be acceptable in principle appears to be dressed up as the argument that we do not have enough knowledge 'yet' to make an acceptable case for its safety. For example, the inspector's report that led to the refusal of permission to the UK waste agency Nirex to build a rock characterization facility (RCF) near Sellafield was based largely on the claimed 'inadequacy' of the science for the project rather than on any *a priori* suggestion that the science could never be adequate. Some anti-nuclear groups commissioned an influential report that concluded that basic research into geological principles would be better value for money than building the RCF. However, the report did not challenge the concept that this research might eventually provide the solution to finding an acceptable site and the technology for deep disposal. But the public rhetoric used by these groups about deep disposal seemed to imply that it would never be possible to attain the necessary certainty. Comparisons were often made between the period of time since the beginnings of recorded human society, some 10,000 years, and the half-life of plutonium-239, which is 24,600 years.

If attitudes towards nuclear power are more fundamental than the arguments used to support them, a reconciliation of the entrenched 'pro' and 'anti' positions is unlikely to be achieved simply by examining the 'facts'. Fortunately, most stakeholders do not operate from the extremes of the debate. The persistence of an acrimonious debate between the small groups of 'theologists' on either side does not necessarily imply the same level of disagreement among the rest of the population about decisions to increase or to reduce the level of nuclear generation.

New approaches to decision-making

There are, as with any other technology, many minor decisions that need to be taken with respect to nuclear power, such as when to return a plant to operation after a planned outage for refuelling and maintenance. It would clearly be impractical for all such issues to be subject to public consultation. Although some pressure groups have recognized that minor decisions are a fertile ground for attacking the viability of the industry, the discussion here focuses on decisions with potentially national implications.

There lies within nuclear technology something of a paradox. The highly technical nature of much of the nuclear field suggests that decisions should be taken by those with suitable experience and qualifications. The social implications of nuclear technology, however, imply that a large number of individuals and interest groups should contribute to those decisions. Accommodating these two requirements in decision-making is a major challenge.

There is no single model of decision-making, even in matters deemed of major national importance. From country to country there can be found a wide range of different governmental structures (and in particular different relationships among national, regional and local tiers), different departmental structures within government (for example, separate energy ministries, energy dealt with by the industry ministry, energy – or some aspects of it – covered by the environment ministry or some combination of these), different relationships with interest groups etc.

Further, decision-making is in practice 'messier' than the models might suggest. In reality there is a large amount of formal and informal feedback. Events intrude and cause delays or fundamental changes in direction. Although governments might like to portray themselves as unified teams, too often they appear as loose confederations of warring tribes, with different departments of state and advisory bodies defending sectional interests. Although 'evaluation' and 'review' of policy are much-praised principles, they are sometimes more difficult to detect in practice.

A number of trends have become apparent in national decision-making in many developed countries in recent years. There has been, as noted, a marked decline in 'deference'. People are much less willing than they once were to accept the word of politicians, scientists or regulatory bodies on controversial issues. New methods of communication, notably the Internet, have made large amounts of information available to individuals who would not have had ready access in the past. Most developed countries are becoming far more pluralistic, with a wider range of world-views stemming, in part, from greater ethnic, cultural and religious diversity. In these circumstances, the scientific dispute over a particular proposal often acts as a proxy for deeper-seated disputes about equity, power relationships and decision-making processes.

In many countries, matters such as site selection for nuclear facilities traditionally followed the procedure that has become known as 'DAD' – decide-announce-defend. Public participation in decision-making, such as it was, tended to be restricted to the final steps in the process. Increasingly, however, this model has failed to produce 'solutions' (as the decision-makers would view it).

In fact, communities that have nuclear installations nearby tend to be among the most pro-nuclear in society. This is undoubtedly because of the direct importance of the facilities to the local economy, coupled perhaps with a diminution of the 'fear of the unknown' and the feeling that there is more involvement in nuclear decision-making than is the case with the wider society. These communities often express resentment at 'outsiders' intruding to campaign against the installation in question.

However, communities in areas without the industry's involvement, and national and international pressure groups, have become increasingly adept at finding an array of measures to prevent major projects going ahead, either on the grounds of NIMBY (not in my back yard) or because of objections to a particular technology for other reasons. These measures include:

- political pressure, including the employment of professional lobbying firms, and especially making use of forthcoming elections, national or local, to create an impression that supporting a nuclear scheme would be unpopular;
- legal and regulatory action, for example using a regional court to overturn federal decisions (where the constitution allows this);
- direct action – blockading construction sites, interfering with transportation routes etc.

Besides having direct effects, these measures create an impression that a new project will be both bad for the image of the companies involved and subject to protracted and costly delay. When most effective, they create a feeling that a major and expensive new project might not even be granted an operating licence when it is completed. This is a powerful deterrent, especially to privately financed operations but also to governments.

In many countries, then, the traditional approach whereby key decision-makers are called upon to decide between two or more 'positions' often proves to be ineffective and expensive. This is especially so in cases involving adversarial scientific and legal processes. The costs, when site security and the effects of long delays are added to planning and legal requirements, can easily run into millions of dollars for major projects. According to one former UK environment minister, 'DAD' has been replaced by 'DADA' – decide-announce-defend-abandon.

The focus of decision-making is therefore now moving more towards earlier and more thorough public participation and focusing on the expectations of various public groups. With respect to a particular major proposal, one can regard society as being made up of three tiers. First, there are people who are already stakeholders – those who clearly have an interest in the decision in question and will make their views known. Second, there are those who will become stakeholders. They are unaware at present that the proposals will affect them 'directly' but will wish to make their views known when they become aware. Third, there are the people who will never be stakeholders, and thus are uninterested in the decision-making process.

It can be argued that the problem with the traditional approach lies with the second group, the 'stakeholders-to-be'. Significant numbers of people become aware only quite late on that a particular proposal will affect them, and by then they are largely excluded from the decision-making process. Often they can make a contribution only when the details of a scheme have been agreed and its proponents are seeking final permission. Public inquiries into proposals have thus become more bitterly fought and more protracted because individuals perceive them as the only meaningful chance to express their concerns or opposition. In effect, they are being asked only to say 'yes' or 'no', and it is perhaps natural that they should tend to the negative.

There is some evidence from countries such as France, Finland and Sweden that offering a reasonable compensation package can to an extent persuade communities of the acceptability of developments such as nuclear projects in their locality. This approach can be particularly effective when coupled with more democratic ways of making decisions.

Increasing the level of public participation in the decision-making process offers two potential benefits. First, a wider range of views can be considered, thus producing 'better' decisions – that is ones that take account of the wishes of a larger number of affected individuals. Second, the process itself will command wider support, so that even those who might disagree with the final decision will feel that their views have properly been taken into account. The involvement of the 'middle tier', the stakeholders-to-be, is especially crucial, and it is becoming increasingly clear that their views and expectations should be sought at a much earlier stage in the process than has traditionally been the case.

Several new consultative techniques have emerged in recent years. Some, notably the 'stakeholder dialogue', focus mainly on parties that are already interested – the first 'tier' described on page 39. (It is, however, quite possible to incorporate awareness-raising into stakeholder dialogues, say by advertising or by organizing appropriate meetings for those involved.) Other techniques seek to involve the wider public – the second and third tiers. Innovative techniques of this kind should be:

- informative – they seek to provide an informed public viewpoint, not instant reactions;
- deliberative – they produce views reached through interactive group discussion;
- independent – they can be independent of the bodies concerned with a final decision;
- inclusive – they seek to involve a wide range of interested parties, including those who are sometimes disenfranchised or underrepresented by traditional approaches.

Techniques for engaging with the wider public include citizens' juries, consensus conferences, interactive panels, deliberative opinion polls and research panels.

There are no rigid models for any of these approaches. The structure of a particular investigation will vary depending on factors such as the kind of proposal involved. A stakeholder dialogue, for example, might involve any number of people, from a relatively small group to several millions, if the people represented by local authorities, trade unions etc. are taken into account. Decisions about how to carry out the dialogue in small, manageable groups, which witnesses to call and how the final report should be written will generally be taken by the particular group of stakeholders involved. The summary of key features of different techniques (Table 2.4), and the more detailed descriptions in Appendix 2.1, should be considered in view of maintaining those techniques' flexibility.

Broadly speaking, most cases mentioned so far have pertained to policy implementation, for example whether a waste repository should be kept open for a period before final sealing, rather than policy formation, that is whether deep disposal is the right approach. However, considering the implementation of policy seems inevitably to lead to a discussion of the basis of the policy itself. People who were unaware that they would become stakeholders while the early stages of the decision-making process were in progress may find the implementation stage to be their first real opportunity to challenge the basis of the underlying policy. An argument is thus emerging that in many cases inclusive decision-making should start at an earlier stage, at the formulation of policy, and that efforts should be made to involve potential stakeholders at that point.

Assessment of new techniques

The assessment of these innovative techniques has reached only a preliminary stage, but some commentators make the following observations concerning:

- *inclusiveness.* No single technique can claim to include all perspectives and social groups. For example, the participants in a consensus conference will be drawn de facto from the section of society whose members would be prepared to give up the necessary time. This section might well not be representative of the population at large. A range of different techniques is needed to ensure that diverse voices and groups are heard;
- *'members of the public', 'stakeholders' and 'experts'.* Some of the techniques require people to be categorized as 'members of the public', a particularly problematic categorization, or as stakeholders or experts. However, these categories are not mutually exclusive. For example, experts can

Table 2.4 Key features of innovative techniques for public involvement in policy formulation

Technique	Number of people/ collective or individual view	Degree of deliberation	Local/ national	Duration/cost	Written briefing/debate	Witnesses cross-examined	Report
Citizens' juries	12–16/ collective	Low	Local	One-off/moderate	Yes/yes	Yes	Yes – by jury
Consensus conference	10–20/ collective	Very high	National	One-off/high	Yes/yes	Yes	Yes – by panel
Interactive panels	12/collective	Low	Local	Continuous/moderate	Yes/yes	No	Yes – by researchers
Deliberative opinion poll	250–600/ individual	High	Local/ national	One-off/very high	Yes/no	Yes	Yes – by organizers
Research panels	500–5,000/ collective or individual	Low	Local/ national	Continuous/moderate	Yes/no	Potentially	Yes – by organizers

also speak as members of the public or as stakeholders. Care must therefore be taken in assessing public, stakeholder and expert views;

- *value of deliberation*. Members of the public must be given the opportunity to respond to issues raised by policy-makers, to challenge experts and to come up with their own unprompted solutions and ideas. Experts are encouraged both to consider how information could be made accessible and to be honest in the face of challenges;
- *degree of deliberation*. This varies from technique to technique. In general, as the number of people involved increases, the amount of deliberation decreases;
- *use of appropriate techniques*. Techniques should be selected as appropriate to the purpose of the consultation. In deciding which techniques are appropriate, it is important to consider matters such as the extent to which participants should represent different social groups, what information should be provided, how much time should be devoted to the process, to what extent participants should be able to scrutinize the information provided, whether participants need to discuss the matter with experts, whether the process should be open or closed and who controls the process and to what extent;
- *integrity*. All parties must have confidence in the integrity of the process, whatever techniques are employed. If a consultation technique is thought to be manipulative or distorting, whether intentionally or not, it will lose its value to the public and to policy-makers.

Using participative techniques in policy formulation

Clearly, the ultimate purpose of the new participative techniques is to influence policy formulation in issues such as the case for building new nuclear reactors and the long-term management of radioactive wastes, as well as a number of issues in other areas of public controversy. In order to do this it will be necessary to create a review process that identifies the nature, purpose and requirements of the key stages in policy formulation. Once again, the caveat should be made that, in practice, decision-making is rarely a smooth process. Further, policy can change significantly during its implementation, either because of changes in the 'environment' in which the decision was first made, for example the development of new technologies, or because of experience of the implementation itself.

Once the general requirements of a particular field of policy have been clearly identified and agreed, one can view policy formulation as comprising five stages:[11]

- identification of public and stakeholder views, concerns and expectations;
- development of a framework for option evaluation;
- provision of an authoritative scientific view of present trends against which to evaluate new options;
- evaluation of options;
- creation of policy recommendations.

A suitable oversight body would need to be created to carry the process forward. The criteria for its suitability might include openness; transparency; a mixture of different interest groups; and inclusion of critics of 'accepted' paradigms of thinking.

A more detailed consideration of how the various techniques described above could be used within the decision-making process is presented in Appendix 2.2.

Volunteer communities

The problem of site selection, for example in developing waste repositories, is especially difficult. In many countries, proposed sites have been rejected after political or legal campaigns. This has led to increasing attention being paid to the 'volunteer community' approach. The development of a major facility such as a waste repository would bring both benefits and disadvantages to a local community. In addition to the natural advantages of improved infrastructure and employment opportunities, a programme of 'planning gain', a series of benefits to the community not directly connected to the development but designed to compensate for the disadvantages (which might include 'planning blight' and perhaps public concerns about safety), could be offered to communities prepared to consider hosting such a facility. As already noted, this approach has been followed with success in a number of countries, notably France. A list of perhaps 10 or 15 areas likely to satisfy technical requirements, including geology and potential transportation links, would be assembled. The communities in question would then be invited to register an interest in hosting the project.

The experience of the 'volunteer community' approach in a number of countries is mixed, but there are indications that success is more likely if local communities have three guarantees. First, they may withdraw from the siting process at any stage prior to the start of construction. Second, the final decision to proceed with repository development may be put to a local refer-

---

[11] RWMAC (2001).

endum. Third, they can participate significantly in the appraisal of site investigations, for example through the setting up and funding of a local advisory
group for ensuring that the concerns of the community are adequately addressed
as investigations progress. This experience also provides valuable lessons about
the measures likely to be necessary both to encourage communities to come forward and to compensate for the blight associated with repository development.

Unresolved issues

It is clear that in many countries, the traditional, centralist approach to making
decisions about matters such as site selection does not work effectively, and, in
some cases, not at all. A move towards more democratic and participative
decision-making would therefore seem to be unavoidable, even leaving aside
matters of equity and fairness.

Perhaps the most acute unresolved issue is the length of time required for
the decision-making processes and the associated question of their costs and
who should bear them. There is suspicion among some commentators that the
whole exercise is simply a delaying tactic, an attempt by those who oppose a
particular technology or project to undermine its economics by introducing
long delays during the planning, and possibly the regulatory, stages.

Planning and/or licensing processes could well be enough to kill off nuclear
investment in a competitive market, whatever its merits. The prospect of a
five-year-plus planning and licensing process attached to nuclear projects is
highly unattractive. There is also the possibility that the plant might lie idle
for several years after completion before operating permission is granted – a
fate that has befallen a number of nuclear facilities. These threats do not hang
over investment in other energy sources such as natural gas, coal or offshore
wind. In a perfect world one would hope that if it were the intention of governments to close down the nuclear option in practical terms, they would be
transparent about it, rather than hiding behind planning procedures.

There is a clear tension between attempts to speed up the planning and licensing process, perhaps through removing some powers from local government
and vesting them in central government, and the desire to involve more stakeholders in decision-making, an inevitably long process. Although the traditional
approach has been singularly unsuccessful in recent years in delivering rapid
and workable decisions, to replace it with a procedure that would result in
enormous delay is not necessarily a good solution.

Participative decision-making techniques demand clarity at the outset
about how the final decision is going to be taken, and by whom. It must be

demonstrated that the whole process has a real effect on decision-making. If stakeholders, or members of the public, devote considerable time and effort to a process at the end of which their views seem to have been ignored, they could become more cynical about the process than if their views had not been sought in the first place. It is particularly important to avoid the impression that the decision has effectively already been taken.

Critics of 'consensus'-based approaches believe that in at least some cases they are unrealistic, and serve merely to cover up many differences and unresolved issues. They argue, for example, that reports can easily be drafted to be acceptable to a host of different readers without actually achieving consensus and that this can increase scepticism, uncertainty and frustration. But advocates of these approaches accept that consensus is not always possible, or even necessarily desirable, if the 'consensus' involves simply an uneasy truce between opposing camps. However, building trust and understanding opposing factions' legitimate concerns can of themselves be a valuable outcome.

There would also be value in providing a forum for discussing and understanding the different scientific opinions garnered by various interest groups and, where possible, developing ways of reaching a degree of consensus. (Often this will involve clarification of the assumptions on which research is carried out.) Many stakeholders are increasingly sophisticated; they are able to gain access to a variety of information sources and even to commission their own research. Excluding them from the commissioning of scientific advice can often fuel controversy at a later stage. One possible outcome of the stakeholder dialogue might be jointly commissioned research, with agreed terms of reference, from a mutually agreed external agency. However, the disparity in access to funds between different stakeholders, for example the nuclear industry and small local pressure groups, may be a difficulty in this respect. Assistance in funding objectors seems to be a valuable move forward.

However, as suggested earlier, some participants in the nuclear debate perceive support or opposition to nuclear power as an issue of fundamental psychology rather than of the pros and cons of the technology itself. If their attitude towards nuclear power is a symbol of deep-seated values, then consensus on its future is unlikely to be reached.

Furthermore, if, for example, a national waste repository is decided to be essential, it cannot be guaranteed that a volunteer community will be found in areas that are geologically suitable. The potential tension between national policy and local self-determination is likely to continue to be a problem for a 'large' technology such as nuclear power that requires considerable state and security infrastructure to support it. Indeed, this is one of the arguments used by opponents against further development of nuclear power. This issue is by no means restricted to nuclear power, and some commentators have observed that in many countries, NIMBY seems to have been replaced by BANANA – build absolutely nothing anywhere near anybody.

There is a final point. In a pluralistic society there will never be complete agreement about major and controversial decisions such as on the further development of nuclear power or the strategy for dealing with radioactive waste. The final decision will require political leadership, and the decision-maker must be clearly accountable to the wider community as well as to stakeholders. Some commentators have drawn a distinction between 'consensus' in the sense of drawing in a wide range of interests and taking their views into account – a clearly positive exercise – and 'consensus' in the sense of allowing a wide range of minority groups consistently to prevent major decisions being taken, which might not be of such benefit to society at large.

Summary

There is little evidence that populations at large in most developed countries are commitedly anti-nuclear. Despite the heated and acrimonious nature of the debate among those who devote their time to such things, most observers and stakeholders seem to take a more balanced view of the issues. It does seem, however, that politicians in a number of countries mistake the heat of the debate for major public disquiet. Polls of politicians in countries such as the United States and the United Kingdom show that they overestimate public opposition by a large margin.

However, proposals in many developed countries to build new nuclear facilities, especially on greenfield sites, provoke considerable opposition. This is true even in countries such as Japan, which, with limited access to alternative sources of energy, have a history of support for nuclear power. This opposition seems to have grown steadily from the inception of nuclear power, and it may be ascribed to a number of factors:

- disillusionment with and mistrust of the use of science (especially science sponsored by commercial concerns) in policy-making. This has been caused in part by widely publicized and exaggerated claims on behalf of technology, notably nuclear technology in its early days, and examples of dishonesty and 'cover-up'. The decline of 'deference' is also relevant;
- poor plant performance, during both construction and operational phases;

- the accidents at Three Mile Island and Chernobyl and other, more localized events;
- campaigns against nuclear power by pressure groups and the media;
- the particular nature of nuclear risks – unfamiliar, involuntary and potentially affecting many people;
- the discovery of major resources of natural gas and oil, reducing the impression that nuclear power would be required in the short term;
- the failure of government and the nuclear industry in most countries to proceed with waste disposal facilities.

It appears that public opinion in some, but by no means all, developing countries may be more favourable to nuclear power. The demand for power is growing rapidly, and local communities often welcome the major employment opportunities afforded by large nuclear construction projects in regions of considerable poverty. But one might speculate that the same forces that have affected public opinion in some developed countries may in due course come to bear in the developing world.

The traditional approach to making decisions about major projects in the nuclear field has been for the industry and government to take the initial steps in private, with little or no public discussion, and then to announce the result of the deliberations and a programme of 'selling' the decision to the public, regulators and planning authorities (the 'DAD' model). Quite apart from legitimate questions about whether this decision-making process is equitable in a complex society, it has become clear that this approach no longer works in many developed countries. Decisions that are taken in secret and that do not reflect the needs of a wide range of stakeholders are increasingly being rejected, for example at the planning stage, because of political lobbying or because of direct action. Such activities tend to drive up project costs, by lengthening the planning phase or requiring more expenditure on site security. This benefits opponents of nuclear energy more than its advocates. Given the age profile of stations in many developed countries, paralysis in decision-making about new nuclear reactors must lead to a significant decline in nuclear output in a few years.

As a result, several new decision-making approaches are being tested in different countries. They aim to improve the quality of the final decisions and to increase confidence in those decisions by fostering a sense of trust among the stakeholders and a wider sense of 'ownership' of the decisions. These approaches share a number of features. There is involvement by a wider range of parties – existing stakeholders, potential stakeholders and members of the wider public – in the decision-making process. This involvement has to start at an early stage of the process. There is greater transparency over the reasons for proposals and also a greater say among possible host communities as to whether, and how, they wish to be involved in discussions.

Many of these approaches are still being evaluated. Although early experience does seem quite promising in some countries, there are major concerns about the length of time they may involve in areas where delay is very expensive. However, it is clear that the traditional approach is no longer an option in many developed countries. The development of new approaches that could be put into practice without causing unmanageable delay seems to be a prerequisite for deciding about replacement nuclear capacity, let alone major expansion, should this be deemed appropriate on other grounds. These issues are by no means unique to the development of nuclear power.

Nonetheless, there is the potential danger that consensus-building might come to be regarded as a substitute for decision-making. Progress on complex issues such as nuclear power, which have scientific, technological, economic, environmental, ethical and social implications, would seem to require one of two things: consensus or political leadership.

They are not mutually exclusive. The greater the degree of consensus, for example, the less the need for political leadership – in the sense of a preparedness on the part of the political establishment to take decisions even though they will inevitably lead to discontent among some groups in society. But it appears likely that full consensus will never emerge over many nuclear issues.

Let us take the example of radioactive waste management. It is unlikely that any action taken today could prevent people using nuclear power in the future if the imperative and desire were sufficiently great. Further, if a workable way forward on radioactive waste were found, then nuclear power would look more attractive, perhaps considerably more attractive, to the general public in a number of countries. But for some groups, opposition to nuclear power is an absolute. It would seem to follow that, for them, it will always appear dangerous to support any measures that might result in a publicly acceptable way forward on waste management, whatever their technical merits or otherwise. And if these groups have some sway with the media and the political establishment, a 'publicly acceptable' route for radioactive waste management may not be possible in principle.

Whatever difficulties they may present, then, the new approaches to democratic decision-making offer a number of advantages over traditional approaches. However, if the attempt to build consensus is interpreted as an alternative to political leadership, then the result could be an avoidance of decision-making against the mythical day when the dispute will have subsided. A proper degree of involvement by a variety of publics into the formulation and implementation of policy is necessary, but someone somewhere will still have to make the decision. In a democracy, it is difficult to see who the decider should be if it is not an elected representative of society at large.

References

Analysis Group (Sweden) (2000), *Poor support for the Government's nuclear power phase-out policy. www.analysisgruppen.org/engopin.*

BNFL (1999), *MORI UK Opinion Poll Findings.* Warrington, Cheshire: BNFL.

IEA (2001), *Nuclear Power in the OECD.* Paris: OECD–IEA.

Lee, T. R., J. Brown, J. Henderson, A. Baillie and J. Fielding (1983), *Psychological Perspectives on Nuclear Energy, Report No 2: Results of Public Attitude Surveys Towards Nuclear Power Stations Conducted in Five Counties of South West England.* London: CEGB.

Marris, C., I. Langford and T. O'Riordan (1996), *Integrating Sociological and Psychological Approaches to Public Perceptions of Environmental Risks: Detailed Results from a Questionnaire Survey,* CSERGE Working Paper GEC 96-07, Centre for Social and Economic Research on Global Environment. Norwich: University of East Anglia.

Oughton, D. H. (2001), *Causing Cancer? Ethical Evaluation of Radiation.* Oslo: Oslo University.

Possony, S. (1955), 'The 'Atoms for Peace' Programme', in F. L. Anderson (panel), *Psychological Aspects of United States Strategy: Source Book,* folder Rockefeller (5), Box 61, White House Central Files, Confidential Files, DDE, p. 203. Washington, DC.

RWMAC (2001), *Advice to Ministers on the Process for Formulation of Future Policy for the Long-term Management of UK Solid Radioactive Waste.* London: RWMAC.

Slovic, P., B. Fischhoff and S. Lichtenstein (1980), 'Facts and Fears; Understanding Perceived Risk' in R. C. Shwing and W. Al Albers (eds), *Societal Risk Assessment: How Safe Is Safe Enough?* New York: Plenum.

Weart, S.R. (1988), *NuclearFear: A History of Images.* Cambridge, Mass.: Harvard University Press.

Worcester, R. (2001), *Public Opinion in Britain and its Impact on the Future of Britain in Europe.* London: Institute of Directors.

Appendix 2.1
Innovative techniques to increase participation in decision-making

It is perhaps natural that the majority of people wishing to influence the nuclear energy debate seem to concentrate on the outcomes of that debate and on whether or not these outcomes bring the world closer to their position. At least until recently, relatively little attention has been paid to the way in which those outcomes are reached. It is important that when seeking to improve decision-making, we make the distinction between the content and the process. Investing too little time and effort into the process itself has repeatedly resulted in decisions that have not been sufficiently based on the issues and on an understanding relevant to stakeholders. Clearly, when affected individuals or groups feel excluded from the decision-making process they also feel less confidence in its outcome. There has been a much-discussed trend, at least in some developed countries, away from trust in the decisions made by elected politicians and towards direct action and local empowerment, sometimes through actions that are themselves unlawful – the so-called 'democratic deficit'.

Several innovative techniques are being developed to allow for greater participation by affected parties. In each case, an attempt is made, in effect, to find a group of 'surrogate stakeholders-to-be'. A group of individuals is identified which, at the moment of asking, represents a balanced cross-section of the population and has no particular interest in a major issue in dispute. The individuals are then given information about it, as a local community would be if it were to find itself the subject of a major proposal, and asked a variety of questions in order to find out how they might react in those circumstances. Although a 'surrogate' group could not be expected to behave in the same way as a real community in such circumstances, it may well carry more credibility with that community than existing stakeholders (pro or anti), who by definition cannot be representative of 'stakeholders-to-be'. (There have been examples of individuals who, having been selected as 'representatives' of the wider public for the purposes, say, of a consensus conference, move on to become active campaigners or stakeholders on the issue in question.)

Advocates argue that although the process can require considerable time, financial cost and commitment from participants, it may well absorb fewer resources overall than more traditional or legalistic approaches to conflict management, environmental strategy-building or general decision-making. These techniques view decision-making as an 'iceberg'. Adversarial processes take place at the top of the iceberg, with issues such as 'build a waste repository at a particular location' versus 'don't build the waste repository'. Each

Figure 2A.1 'Iceberg' diagram of decision-making

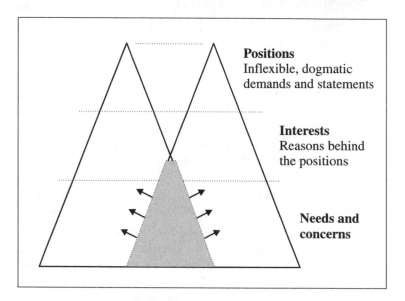

side, it is argued, only guesses what motivates the other, and makes little or no effort to investigate the interests and concerns underlying the position statement. The new techniques seek to find common ground (shaded area in Figure 2A.1) and develop understanding of the interests of each party from that firm footing (arrows outwards and upwards).

Advocates argue that this is the key to constructive dialogue. Positions are the 'tips' of icebergs, informed both by the interests and, more fundamentally, by the needs of the parties concerned. The further one goes beneath the surface, the more the interests and needs of the parties are likely to overlap – all sides, for instance, may want a healthy environment for their children or local jobs for young people.

A distinction can be drawn between processes that seek to gain better understanding between those who are already aware that they are 'stakeholders', and those which seek to involve people who either do not yet realise that they may be stakeholders, and those who are unlikely ever to become stakeholders. (One might observe, however, that nuclear power has the potential to affect very large numbers of people indeed, both in its positive aspects such as energy production and greenhouse gas emission mitigation, and in its negative features such as waste management or the threat of major accidents.)

It should also be stressed that the following descriptions are merely examples: the innovative techniques are flexible, and an investigation of expectations associated with any particular field or proposal would be tailored to its specific requirements.

Perhaps the main technique for involving existing stakeholders is the *stakeholder dialogue*. Examples of stakeholders might include:

- people working for the organization funding the project, including both managers and representatives of the workforce;
- members of expert groups, such as learned societies, consultants and academics;
- government;
- regulators;
- members of representative groups, such as elected politicians (local or national);
- environmental or other pressure groups (local or national);
- consumer groups;
- foreign countries;
- interested individual citizens.

A stakeholder dialogue will typically be overseen by an independent and credible third party. It will start with process design by the project facilitation team – those who have pioneered this approach believe that the quality of preparation is a crucial element in the success or otherwise of the whole process. Stages in the process may include:

- identification of a range of possible participants;
- agreement over details such as an outline agenda, draft ground rules, composition of teams, times and venue;
- the bringing together of representatives from all sides in a facilitated meeting;
- one-to-one meetings with citizens, focus or work groups;
- visits and help from special advisers, ensuring that adequate information is available to the participants;
- jointly commissioned surveys or other research.

All stages are driven by a desire to ensure effective communication among all stakeholders and, when appropriate, with those outside the dialogue process. A key principle is that as much of the process as possible should be decided by the stakeholders.

Outcomes can be many and varied, visible and invisible (such as increased trust between stakeholder groups, which may provide a platform for more constructive negotiations on future potential conflicts). The visible outcomes may take the form of the following:

- key recommendations to a company or trade association dealing with a particular issue;
- consensus among participants on the implementation of a plan or adoption of best-practice guidelines;
- a jointly agreed response to government consultation on a certain issue;
- a long-term strategy supported by stakeholders who will be interested in the issues in the years to come;
- sustainability indicators for sectors/organizations;
- social inclusion in decision-making at a local level regarding local infrastructure, local policy and the like.

A few examples of new techniques used to elicit the views and involvement of people who are not (yet) stakeholders in the above sense are described below. They are meant to be complementary, rather than any one being sufficient. In general, there is a trade-off between the number of people involved and the depth to which their views can be developed and examined. As noted earlier, there is also an issue about the selection of participants. For example, techniques that require a significant time commitment from the participant will only attract people who are prepared to make that commitment. The extent to which they represent the wider community will inevitably be questionable.

Citizens' juries aim to involve the public in decisions that affect them in their own communities. A group of local residents (say 12 to 15 people) is chosen to be broadly representative of the area. They then sit for up to four days to consider policy proposals and local impacts (assisted by independent moderators); they cross-examine witnesses and compile a report and submission to the commissioning body. Proponents claim that citizens' juries have a unique combination of characteristics:

- they involve ordinary members of the public in their capacity as citizens;
- they ask jurors to take part in a 'serious civic task';
- they provide an open democratic mechanism to complement existing bodies and forms of public consultation;
- they allow issues to be considered in detail.

Consensus conferences involve a panel of lay people who develop their understanding of technical or scientific issues in dialogue with experts. This approach was developed in Denmark, and has been used in countries such as the United States and the United Kingdom. A consensus conference involves a panel of between 10 and 20 lay people recruited through direct approach or through advertisement and chosen to reflect the demographic balance of the population as closely as possible. A steering committee, whose members are chosen by the sponsors of the conference, is also assembled. The panel attends preparatory weekends where they are briefed on the subject. The conference itself lasts for three or four days and gives the panel a chance to ask experts any outstanding questions. It is open to the public, and the audience also has the opportunity to ask questions. The panel then retires and, after facilitated discussion, prepares a report. Proponents claim that consensus conferences allow panel members to control the content of the process and seek consensus, and emphasize that they are open events.

Interactive panels have a standing membership that meets regularly to deliberate on issues. Many nuclear sites have had standing site liaison committees, for example, for some years. They have the following features:

- membership composed of perhaps 12 members of the ordinary public;
- recruitment designed to ensure an appropriate range of demographic features;
- regular turnover of members to bring in new voices;
- in some cases, further panels discussing the same issues, to provide more confidence in the outcomes;
- regular meetings, say three times a year or so, to discuss topics set by the panel or a central authority;
- facilitation by an independent researcher;
- votes on each issue at the end of the discussion, providing quantitative information on members' views;
- transcriptions of meetings to provide records of discussions and views;
- preparation by the researchers of a report for the commissioning body.

Deliberative opinion polls measure informed opinion on an issue. The difficulties with ordinary opinion polls intended to 'reflect' the public's views on subjects about which they might know little have been discussed earlier. A deliberative poll is carried out after respondents have had an opportunity to consider the matter in question more closely. The sample – perhaps between 200 and 600 people – is recruited to represent the population in terms of demography and 'base-line' opinion. Written briefing material is provided, followed by two to four days of deliberation in smaller groups and preparation of questions to put to experts in plenary discussion groups. The views on particular questions are measured before and after the process. Supporters of the technique argue that deliberative opinion polls provide a scientific random sample of a population and therefore represent the larger group in terms of attitudes and demographics. They can give an indication to decision-makers and the media about what the public might think if they had enough time and information to consider views carefully. It brings together people from diverse social backgrounds, enables participants to control the agenda and demonstrates how collective views might change after informed deliberation.

Research panels are relatively large samples of the public that can be used flexibly to track changes in opinion over time by way of a number of different techniques. They consist of from 500 to 5,000 participants, recruited either through the post or by telephone. There is a standing membership, but with a proportion replaced regularly. Participants are asked about a range of different issues over a period of time. Advocates argue that a panel is large enough to be representative and offers a wider public view on specific issues. Panel members develop an informed public view. Once the panel has been set up, it is cheaper than a series of large one-off opinion surveys. Panels provide a flexible methodology, as the same panel can be surveyed using questionnaire techniques, or it can be used to seek a more informed view on some issues. Members of the panel can also be brought together in focus groups, workshops, citizens' juries or consensus conferences.

Appendix 2.2
Using participative techniques in policy formulation

It is a considerable challenge to ensure appropriate input to decision-making from the new techniques described in Appendix 2.1 in a way that does not so slow the process that decisions are never taken.

Below is one possible approach, which would allow for the deliberations arising from a variety of innovative techniques to enter the decision-making process.[1] However, it is a matter of speculation as to how long the whole process would take. This approach may be more useful in the case of projects such as a waste repository, in which decisions are unlikely to be required in a short time, than in the case of new nuclear power stations, in which delay can be very expensive.

Once the overall requirements of a particular field of policy have been identified, policy formulation can begin. It has five stages:

1. identification of public and stakeholder views, concerns and expectations;
2. development of a framework for option evaluation;
3. provision of an authoritative scientific baseline for option evaluation;
4. evaluation of options;
5. formulation of policy recommendations.

1 Identification of public and stakeholder views, concerns and expectations

This stage is sometimes referred to as 'front-end consultation'. Its purpose is to:

* understand public views (as noted earlier, what organizations, including governments, think the public believe and what the public actually believe can be very different);
* identify difficult or controversial issues (so that they can be addressed in subsequent stages);
* identify public and stakeholder concerns and expectations (for example, on the balance between national interest and local impacts or concerning intergenerational equity).

It can be helpful to regard this stage as an initial attempt to elicit the views of three groups of people – stakeholders, stakeholders-to-be and never-to-be-stakeholders. However, by definition the second group, stakeholders-to-be,

---

[1] RWMAC (2001).

cannot be polled directly (there are no firm proposals as yet that might cause them to move from non-stakeholder to stakeholder status). This group can be simulated, to a degree, by taking a balanced cross-section of non-stakeholders and providing them with 'tutoring'. A number of issues emerge at this point.

First, as already mentioned, the level of factual knowledge about nuclear power, and the context in which the industry operates, among the general population – that is non-stakeholders, people who do not have a particular reason to be interested – is low. This does not appear to be because the public finds the issues impossible to understand but simply because they perceive little need to make the effort. Once there is a reason for them to become involved, either because they are asked to take part in a consultation exercise or because a scheme is proposed in their vicinity, then they are likely to start gathering information, and their initial view, if indeed they had one, may change. This group is most likely to be close in behaviour and attitudes to the 'stakeholders-to-be', although the view even of a 'tutored' person might be quite different when they were taking part in a consultation exercise from when they were faced with a real proposal affecting them directly.

Second, the nature of the briefing materials for the group to be tutored will be of great importance. Every effort will have to be taken to make sure that the information is factually correct and balanced, giving a fair assessment of the views of different participants in the debate and the reasoning behind them. There should be no attempt to adjudicate between the various views at this stage. Further, the scope of the information will be important – it is hard to consider nuclear waste management policy, for example, without some background in the wider nuclear debate, including patterns of energy demand and the climate change debate.

The techniques used to reach these different groups might include:

- *for never-to-be-stakeholders*: seeking to consult a large number of people, through newspaper ads, free telephone lines, a web site or even traditional public opinion polls;
- *for surrogate stakeholders-to-be*: providing a representative sample of the population with a degree of tutoring, then consulting them using a research panel, telephone surveys, focus groups or a deliberative opinion poll;
- *for stakeholders*: consulting through meetings, workshops or facilitated dialogue.

A summary of the findings of this stage would be made available for public comment in order to help ensure that public views, concerns and expectations have been properly captured for feeding through to the next stage.

2   Development of a framework for option evaluation

The purpose of this stage would be to:

- formulate objectives (in the light of public and stakeholder views, concerns and expectations);
- generate policy options;
- specify criteria (that will be used to evaluate the options);
- identify future scenarios (to explore long-term impacts);
- identify research needs.

Public consultation and deliberation, for example through a consensus conference, would be essential at this stage. There would be a need for effective expert input, such as on the technical feasibility of different options. There would also be value in organizing parallel stakeholder dialogue processes. These might seek to involve all major stakeholders in one facilitated event or entail a series of meetings with groups of stakeholders.

3   Provision of an authoritative scientific baseline for option evaluation

This step is particularly challenging. Controversial scientific fields are characterized by research commissioned by both 'sides' of the debate, often starting from different assumptions and thereby producing apparently contradictory conclusions.

Research relating to a number of options would therefore be required at this stage, perhaps entailing jointly commissioned work. A clear statement of the assumptions underlying the work should be made and, if possible, agreed. An independent multi-disciplinary panel might be convened to:

- systematically address uncertainties;
- identify significant areas of agreement and disagreement;
- identify how uncertainties and disagreements could be reduced and over what timescales;
- seek peer review;

- open emerging findings to public and stakeholder comment at an early stage.

An oversight body would be responsible for reviewing the outcome of panel appraisals relating to all the options under consideration. It would also be responsible for ensuring that these findings are utilized in option evaluation.

4   Evaluation of options

The purpose of this stage would be to utilize a clear and transparent methodology to:

- assess the performance of each option against agreed criteria (including a systematic comparison of risks);
- weight the criteria in order of relative importance;
- perform sensitivity analysis;
- produce an overall ranking of options.

As in the 'framework development' stage, public and stakeholder consultation entailing intensive deliberation would be required. This might be provided by reconvening the consensus conferences and the stakeholder dialogue processes. Consensus conference and stakeholder dialogue reports would then be made widely available for public comment.

5   Formulation of policy recommendations

The oversight body would review the option evaluation reports and public comment. It would then proceed to formulate policy recommendations. Prior to finalizing these recommendations, the oversight body might consider it appropriate to undertake further telephone surveys, focus group discussions or a deliberative opinion poll. Following an appraisal of any further public consultation, the oversight body would finalize its recommendations.

3 The relative economics of nuclear power

Introduction

Society makes several requirements of power generation systems. Perhaps the most important are:

- secure supplies of electricity;
- cheap supplies of electricity;
- environmentally acceptable supplies of energy;
- socially acceptable supplies of electricity, that is safe, not detrimental to health, and consistent with the desire to protect certain communities associated with one fuel or another.

These demands can often be in conflict. Ensuring the safety of power plants against internal or external threats can be expensive; protecting a strategic source of energy such as coal can have adverse environmental implications. The relative weight that governments and industry place on these factors will change from place to place and time to time, depending on prevailing circumstances and on political principle or fashion.

Decisions on investment in power supplies, in developed countries in particular, have increasingly been influenced by economic factors in recent years, especially as perceptions of shortages of hydrocarbon fuels (oil and gas) have receded. The nuclear reactors that have been built in recent years have had capital costs significantly higher than those of fossil fuel power stations. This has rendered such designs economically unattractive for new build in liberalized markets, where shorter payback periods and higher rates of return prevail, than in more centralized power markets.

The relative economics of nuclear power in a competitive market place are very heavily influenced by the rate of return demanded on capital, by the success of the construction programme (in terms of both cost and timetable) and by the reliability of the plant. Fuel costs, and waste and decommissioning costs, are less crucial to the costs of power from a nuclear plant averaged over the plant's lifetime. (Rising back-end costs are important, however, if the rises occur late in the operating lifetime of a particular power station.)

A new generation of power plants with lower initial costs is at an advanced stage of design, and three such designs, the AP600, System 80+ and ABWR, have received regulatory approval in the United States. As yet, none of these plants has been constructed, making firm predictions about their economic performance difficult. Two reactors of an advanced version of the boiling water reactor (BWR) have been constructed in Japan, with two more units under construction in Chinese Taipei (Taiwan), but in the absence of a proven, economically attractive product, the nuclear debate insofar as it covers areas such as waste management, the safety of reactors, the health effects of low levels of radiation, weapons proliferation etc. is likely to be largely irrelevant.

However, two broad developments might change this situation. The relative economics of nuclear power might improve compared with other fuels (as previously noted) because of improvements in nuclear economics or the deterioration of the economics of its main competitors. For example, it is clear that world reserves of hydrocarbon fuels are not infinite. Their physical availability will start to decline at some point, probably between 2020 and 2060, with possible consequences for their relative economics. The period 1999–2001 was characterized by significant rises in global fossil fuel prices, although for reasons of politics rather than because of a shortage of reserves. Alternatively, or perhaps concurrently, there might be a reversal of liberalization of power markets. This might lead to a lengthening of payback periods and a reduction in the demanded rates of return (as presumably there would be less economic risk in a planned system with captive customers). It could act in nuclear power's favour, although, as noted later, there have also been benefits for nuclear power in liberalization.

In principle, the same factors are important both in the developed and the developing worlds. However, the liberalization of electricity supply markets is less advanced in a number of developing countries. In the biggest of them, large nuclear plants remain relatively attractive, especially in the cases where fossil fuel supplies are limited by lack of availability or by the large distances between the fuel reserves and the centres of economic growth. However, in these countries as elsewhere, the higher capital investment costs of existing nuclear designs represent an obstacle to investment.

Even within a liberalized electricity supply market, it is generally agreed that government still has important roles to carry out. These include environmental protection, application of health and safety legislation and prevention of unfair competition. It can also be argued that government has an important duty to ensure regulatory stability, insofar as this can be offered. Unstable regulatory regimes are especially damaging to sources of electricity that have long payback periods, such as nuclear power and renewables. However, there is a vigorous debate over what role, if any, governments should play if security of power supplies were compromised in a competitive market. All of these points are of relevance when comparing the economic attractiveness of different sources of electricity.

This chapter focuses on the use of nuclear power for generating electricity for distribution to industry or domestic users. However, other potential uses of nuclear power, for example to provide the energy to desalinate sea water, to extract oil from more difficult sources such as shale or to produce hydrogen for use in fuel cells, may ultimately become important as well.

The costs of generating electricity

Electricity demand is expected to rise, although not as fast as GDP in some developed countries, and for many of its uses there is no attractive alternative. Not only is world energy demand expected to grow, the proportion of that energy provided in the form of electricity is rising too. Demand in developed countries is expected to increase by perhaps 40 or 50 per cent by 2050, while that in developing countries could increase fourfold or fivefold in the same period.

It is not the aim of this chapter to evaluate the claims made for the costs of various technologies or plants – this must be done by potential investors on a project-by-project basis. In any case, the relative economics of one source of energy in comparison to others is a complex concept.

The average costs per unit of electrical output of major projects are derived as follows. 'Substantive' costs represent out-of-pocket expenditure on real resources used in operations such as fuel production and transportation, power plant construction and electricity generation and distribution. They consist of costs of labour, materials etc. 'Levelized' costs emerge after applying conventions (often financial and especially those determining appropriate rates of return and discounting) to translate expectations of substantive costs and the timing of income and expenditure streams into costs as perceived by investors, that is the net present value of the investment decision. Thus changes, such as in risk perceptions and discount rates, can mean that a project with the same substantive costs can have very different 'levelized' costs, depending on market rules, regulatory regimes, government policies etc.

It is very difficult to cite a single figure for the costs of electricity from a particular fuel. The substantive costs (per unit output) of different coal-fired power projects may vary considerably, depending on the availability of cooling water, the distance from the source of coal, the distance from the major market for the electrical output etc. In China, for example, the enormous distances between the major coalfields in the north of the country and the centres of economic expansion in the south and southeast add significant infrastructure and transportation costs to coal-fired power generation in comparison to localities in which the coalfields are closer to the market place. Similar factors can affect the economics of gas-fired plants. Different designs of nuclear power station have different unit costs; those countries that have replicated a small number of designs have tended to enjoy lower costs than those that have built a series of unique stations. The 'levelized' costs can also vary widely, depending on the structure of the electricity supply market, political support from government, perceptions of economic risk among investors and other factors.

For nuclear power in recent years, there have been problems caused both by increases in substantive costs and by changes in the financial conventions being applied, notably demands for higher rates of return, for reasons discussed later in the section on market liberalization.

In the case of renewables, the variations in costs may be even more marked. The unit costs might be expected to fall more rapidly as mass production is introduced, and one might expect steeper learning curves. This being said, presumably the sites with the best wind cover, the greatest tidal range or wave density and the highest average levels of insolation will be used first, and later units would be faced with more difficult operating conditions.

Current comparative electricity generation costs

As noted on the previous page, it is difficult to place a single value on the costs of any energy source, even given similar regulatory and financial regimes. The Organization for Economic Cooperation and Development's Nuclear Energy Agency (NEA) and International Energy Agency periodically collect data on levelized power production costs from a range of countries, the latest study being published in 1998 (based on 1996 data).[1] This study assumed certain parameters – 40-year plant lifetimes, 75 per cent output factors and fuel costs based on 1996 prices and increasing in real terms by an average of 0.3 per cent per annum for coal and an average 0.8 per cent per annum for gas.

Many of these assumptions have been invalidated by events. Thermal plant output factors now exceed 80 per cent in most cases. (Although this benefits all power sources, it has a disproportionately beneficial effect for nuclear

---

[1] OECD–NEA (1998).

power, which has relatively low marginal costs.) Oil and gas prices have risen significantly. The coal price in the western American coalfields doubled in 2000–1 because of increased power station demand. The reported sharp drop in Chinese coal production in the late 1990s raised questions about assumptions of plentiful cheap coal in that country.

Furthermore, the quoted figures are based on 1996 exchange rates, some of which have changed significantly in the meanwhile.

The 'precise' figures for energy sources must therefore be treated with considerable caution. However, they can still be used to illustrate the enormous significance of changing discount rates on the relative economics of different options (see Table 3.1).

As can be seen, the effect of raising the discount rate is least serious for gas-generated electricity and most serious for nuclear-generated electricity. The two quoted discount rates of 5 per cent and 10 per cent were chosen for comparison purposes in the OECD–NEA study. Competitive commercial markets may well demand rates in the region of 15 per cent, in which case traditional nuclear plants would, all else being equal, appear to be even more uneconomic in comparison to fossil fuel alternatives.

There is also likely to be a greater range of uncertainty in the projected costs of power production from nuclear power in several countries than in the corresponding costs from fossil fuels. Many countries have recent experience of constructing gas-fired power plants, and thus have direct experience of actual costs. This is not the case in the nuclear field, with the exceptions of France and some countries of the Pacific Rim. Evidence from the latter seems to suggest that the costs of nuclear construction projects are under more control now than was the case 10 or 20 years ago. This applies as well to construction projects involving other fuels, as project management efficiency in general has

Table 3.1 Total levelized costs* for different fuels, 1996: the average of five selected countries**

Source	Costs at 5% discount	Costs at 10% discount	% increase in levelized costs on moving from 5% to 10%
Gas	4.5	4.9	10
Coal	4.1	5.3	30
Nuclear	3.7	5.6	50

* 1996 ¢ per kWh.
** Finland, France, Japan, Russia and the United States.
Source: OECD–NEA (1998).

increased, but it may be a more important factor in the case of nuclear plants, as the construction phase accounts for a larger proportion of total costs.

The economics of renewable forms of energy can be even more difficult to express in a single figure, and there is relatively little actual experience, certainly in comparison with gas or coal-generated power plants. Further, some of the key parameters used in the OECD–NEA study are meaningless with reference to some renewables, notably the load factor of 75 per cent. However, it appears that the generating costs of renewable forms of energy are generally falling at a more rapid rate than alternatives. The rate of innovation is higher for relatively new technologies, and there are considerable economies of scale and benefits from greater operating experience to be enjoyed in rapidly growing industries.

In much of Europe the major wind resource is offshore. The United Kingdom's first offshore wind project, opened recently and using 2-MW machines, is said to be generating at less than 7¢ per kWh (discount rate six to seven per cent).[2] Experience elsewhere in Europe suggests that offshore wind prices can fall further. Denmark plans to develop 750 MW of offshore wind power by 2008, with a further target of 4,000 MW planned for 2030. It is claimed that costs could be between 4.3¢ to 4.7¢ per kWh. The Dutch government has plans to install 3,000 MW of wind power capacity by 2020, half of it located offshore; the costs of electricity are expected to be between 3.3¢ and 6.6¢ per kWh.

However, should wind generate more than about 15 to 20 per cent of electricity demand, problems would arise over integrating the intermittent output in such a way as to guarantee secure supplies. In any case, back-up capacity will have to be available to cover variations in wind cover. At present, wind generators operate at approximately 30 per cent availability, and the power is not predictable. To safeguard security of supply, one would need either storage capacity or back-up generating capacity (probably gas-fired) to cover the 70 per cent of rated output when windpower was not available. The capital and operation and maintenance costs of this back-up must be covered for the 30 per cent of its potential output which is displaced by windpower. Whether or not the costs are charged to the windpower operators or to elsewhere in the system, these costs are a result of the intermittent nature of the power source itself. These 'spinning-out' factors would therefore increase the effective costs of wind-powered electricity generation. The same applies, to a lesser extent, to other intermittent renewables that are more predictable, such as solar power and tidal power.

---

[2] Elliott (2001).

The costs of wave and tidal current power also seem to be falling. LIMPET, the world's first commercial grid-linked wave device, situated on the Isle of Islay in Scotland, is generating at about 8.5¢ per kWh. In a number of countries, at least some consumers are prepared to pay a premium for electricity generated using renewables; the size of this premium has been relatively modest. In a sense this behaviour is economically irrational (in that the final product is identical, however it is generated), but it reflects their willingness to pay more in order to protect the environment. A similar phenomenon is the growth of 'ethical' stock market funds, which may act as an extra source of capital for renewable energy schemes. Although difficult to model, consumer preference may have a direct effect on the relative economics of different energy sources.

In any case, cost comparisons between renewables and nuclear power may be of limited relevance, at least for the present. Nuclear power is used for baseload power production; the intermittent nature of some of the new renewables, notably wind and solar, makes them unsuitable for baseload production. (Tidal power is predictable, although the output of a tidal barrage varies significantly with the point in the tidal cycle.) However, the development of large-scale methods of storing electricity directly or indirectly, for example hydrogen for fuel cells or direct use, would change this situation. The dream of finding a method of storing electricity is a very old one, and it is not clear that a solution is any closer today than it was some decades ago.

Recent trends in power production costs

Nuclear economics have broadly improved in substantive terms in the past two decades, especially because of the enhanced output reliability seen in many countries. The capacity factors of the world's pressurized water reactors (PWRs) and boiling water reactors (BWRs), which together have accounted for between 85 per cent and 90 per cent of world nuclear output, have changed, as set out in Table 3.2.

Table 3.2 Global capacity factors (%) for light water reactors, 1980–2000

	1980	1985	1990	1995	2000
Capacity factor (PWR and BWR)	58.6	68.7	68.8	75.6	80.1

Source: *Nuclear Engineering International* (1981, 1986, 1991, 1996, 2001).

However, nuclear costs have not fallen as rapidly as have those of natural gas, especially since the development of the combined cycle gas turbine (CCGT). Between the 1992 and 1998 OECD surveys of levelized costs of electricity production in a variety of countries, the costs (corrected for inflation) of electricity generated using gas fell between 16 per cent and 54 per cent. The costs of electricity made using coal fell by between 3 per cent and 34 per cent, and those of nuclear power by between 2 per cent and 27 per cent.[3] The substantive costs of several renewable options have also been improving more rapidly than those of nuclear power, albeit from a higher base.

A number of reasons have been adduced for the decline in the relative economic position of nuclear technology. This technology has not changed fundamentally since the 1960s and 1970s (when the majority of plants now operating were ordered). In many cases, the record of project management during the construction and commissioning phase of nuclear plants was poor. By contrast, the CCGT, which has had a significant effect on the thermal efficiency of generating electricity from natural gas, was made possible by the development of turbines for military jet engines, and took hold largely as a result of the liberalization of the late 1980s and the 1990s.

Furthermore, nuclear power has operated under an increasingly tight regulatory regime in most developed countries, in part owing to the accidents at Three Mile Island in 1979 and, possibly to a lesser extent, to Chernobyl in 1986. (Three Mile Island, although it caused no off-site health consequences as far as can reasonably be determined, occurred in a US-designed pressurized water reactor, similar in concept to those generating the majority of the world's nuclear capacity. Chernobyl, although far more serious in its consequences, resulted from the failure of a water-cooled, graphite-moderated reactor, a design that was built only in the former Soviet Union and that would not have satisfied Western safety standards. It can therefore be argued that this accident has less relevance for reactor systems elsewhere.)

The features of this regulatory regime include:

- a tightening of worker-dose limits, following revisions in the dose-response curves adopted by the International Commission on Radiological Protection. Average annual doses have decreased by approximately a factor of four over the past decade, requiring different working practices and investment in plant in order to achieve reductions in worker doses;

---

[3] OECD–NEA (1998).

- increasing public awareness of nuclear discharges and increasing demands to treat waste and discharge streams with ever more stringent standards, even where, in the view of the nuclear industry, cost-benefit analysis would not support the tightening of standards;
- continual tightening of emission and dose standards by the regulators, as a response to increased public and political pressure;
- a trend for plants to have extra safety measures backfitted, rather than incorporated as an integral part of the design, generally because of a change in standards of construction or operation – this was especially prevalent among plants being constructed at the time of Three Mile Island.

The nuclear industry argues that the relentless external pressure for zero discharges and exposures, even where there is no technical case for it, has damaged nuclear economics while having little effect on the environment or human safety. The industry contends that in many ways it has been pushed to adopt standards that are far more stringent than in other industries. In part, it is suggested, this is because the industry's activities are emotive and politicians perceive the need to be seen to react to stakeholders' concerns. In addition, the detection thresholds for radioactivity are very low. There is a common perception that the very fact that a discharge is measurable, even if it is totally insignificant radiologically, must make that discharge practice unacceptable.

Those opposing nuclear technology argue that radiation is dangerous, even at very low exposure rates (often expressed in the phrase 'there is no safe dose of radiation'), and that every reappraisal of the risks of radiation has concluded that it is more dangerous than previously calculated. They therefore consider it entirely appropriate that regulation should continue to tighten.

Another factor has been the rising costs of the back-end of the nuclear fuel cycle – waste disposal, decommissioning and reprocessing – in many countries. Technically this does not affect calculations of the levelized costs of plants to be commissioned between 2005 and 2010, as these costs include estimates for the back-end costs based on current best assessments. Further, as these costs are incurred late in the investment cycle for the project – after the closure of the station in many cases – their discounted values account for a small proportion of the total levelized costs – typically less than 10 per cent.

These cost rises have nonetheless been important, in two main ways. First, back-end cost increases, which occur after a reactor has been operating for some years, cannot easily be accommodated, as the period of time over which the extra funds can be raised (before the station stops generating electricity and thus an income) is short. In a competitive market it is in any case unlikely that prices could be raised to compensate. Second, the impression that back-end costs may continue to rise adds an extra degree of economic risk to new nuclear projects. This effect could be offset, at least to an extent, by the trend towards extending the life of reactors.

There is a degree of agreement that regulation should be based on an objective assessment of the risks involved. However, until the scientific community concurs on quantifying the risks involved, disagreements over the appropriate level of regulation will persist. One obvious example is the possible threat of terrorist attack on a nuclear power station (especially after the events of 11 September 2001), which involves several uncertainties, including the attractiveness of this target, the technical means at the disposal of the terrorists and the effects an attack might have on the various designs of power station.

Furthermore, risk assessment must be complemented by a recognition of the particular public concerns associated with radiation, and an approach must be established to ensure that these concerns are taken into account (an issue considered in more detail in Chapter 5, on nuclear safety, and Chapter 2, on decision-making).

Financial conventions have also turned strongly against nuclear power in some developed countries. As discussed later, liberalization of electricity supply markets has profoundly altered perceptions of economic risk, and has demanded rates of return associated with investment in power plants in a way that has been largely unhelpful to the nuclear industry, and to renewables as well. (By contrast, of course, calculations of levelized costs could in principle turn in nuclear power's favour again if the financial conventions themselves changed, for example if carbon taxes or tradable emission permits were introduced as part of a policy aimed at reducing emissions of greenhouse gases.)

However, some of the above factors apply to other sources of electricity too. For example, stricter controls on emissions of acid rain gases, notably sulphur dioxide, have had a significant effect on the economics of coal-fired power plants, either through the backfitting of flue gas desulphurization (FGD) equipment or through 'clean coal' technologies with higher installation costs. Should sequestration of carbon dioxide from the flue gases of power plants become mandatory, this could add between 40 per cent and 100 per cent to power costs from fossil fuels. (But it should be noted that the relevant technologies are relatively new, and large cost savings may be possible as they mature.[4])

---

[4] Audus and Freund (1997).

Similarly, although the cost of generation equipment for renewables is likely to fall, this may be balanced to some extent by an increase in other costs. For example, in the case of wind, there may be difficulty in obtaining planning permission for sites with good wind cover, and in any case the economics of using sites with poorer wind characteristics will be less favourable than those of the best sites. Further, as noted earlier, should use of the intermittent renewables expand, it will be necessary to maintain back-up capacity for the times when the plant is not operating, and this will affect the overall economics of the system.

A further observation is that in the field of economics, as in other areas of the nuclear debate, perceptions and fashions can change rapidly, making 'trends' difficult to project. Nuclear power fell from favour quite quickly in many countries in the 1980s. There were several reasons for this – overruns in investment costs, poor operating performance, public concerns and overcapacity in electricity production in many countries – and the liberalization of markets has exacerbated the issue in relation to new construction. However, the situation in the United States has recently been moving just as rapidly in the opposite direction. Just two or three years ago it was assumed that many nuclear stations would be closed before the end of their design lives, and the market price for existing plants was practically zero. Now there is a highly competitive market for operating nuclear stations, and it is assumed that at least 80 per cent of the stations will seek extensions to their lifetime. The reasons for the change are several, including a significantly improving output performance, an increasing shortage of generating capacity in several areas of the United States and concerns about the possibility of future rising prices and limited availability of natural gas.

The expenditure profiles of different sources of electricity

In a competitive market, projections of levelized costs over the lifetime of a power project are only one consideration. The point in time at which investment is required is also centrally important. The earlier the investment has to be made, the longer the payback period and the greater the economic risk in the absence of long-term contracts for the output. As can be seen in Table 3.3, capital costs dominate the economics of nuclear power, accounting for some 70 per cent of total costs even at a 10 per cent discount rate. This can be contrasted with other sources of electricity (also at 10 per cent discount rates).

The fuel component of the nuclear cycle consists of three elements in roughly equal proportion: mining uranium, refining and enriching it; processing uranium into fuel elements; and dealing with the waste. The actual costs of

Table 3.3 Distribution (%, rounded) of costs of generating electricity at 10% discount rate

Source	Investment	Operations and maintenance	Fuel
Coal	48	16	36
Natural gas	27	9	64
Nuclear	72	17	12

Source: OECD–NEA (1998).

the raw material represent only about two per cent of total cycle costs. Nuclear electricity is therefore extremely resistant to fuel price inflation, although by the same token it benefits little in times of falling fuel prices.

Nuclear economics could be affected in a number of ways by an event like the terrorist attacks of 11 September 2001 and subsequent concern that nuclear power stations might become terrorist targets. For example, the capital costs of construction could rise because of extra safety measures such as advanced outer containment, and the operational costs could rise because of the need for extra security. Moreover, appraisal conventions could change if nuclear power were regarded as a more risky investment than previously, resulting in higher demanded rates of return.

The distribution of the costs of renewable forms of electricity production are typically similar to those of nuclear power, with high investment costs (especially for initial construction) but fuel costs zero in a number of cases. However, decommissioning costs are likely to be far lower than those of nuclear power, and in many cases there are no ongoing waste management costs.

Total investment costs for new nuclear capacity, for designs that have been constructed in recent years, range between $2,000 and $2,500 per kW(e). (Westinghouse claims that the installation costs for the AP600 may be about $1,500 per kW(e) and that the costs for installing the AP1000 may be perhaps as low as $1,000 per kW(e), although this will not be verifiable until demonstration units have been constructed. The Canadian company AECL claims capital costs of about $1,200 per kW(e) for the Next Generation CANDU reactor.) For coal-fired power stations the range tends to be between $1,000 and $2,000 per kW(e), depending in part on the measures included to reduce the emission of sulphur dioxide and particulates. For CCGTs the capital costs range between $500 and $900.

The distribution of costs between investment and running costs (operation and maintenance (O&M) plus fuel) confers certain characteristics on the economics of various fuel sources. As an example, the costs of nuclear power tend to be relatively insensitive to changes in the fuel price, as noted. In other words, nuclear power is inflation-proof against rising fuel prices, although it gains little benefit when fuel prices are falling. However, it is highly sensitive to matters affecting the investment segment of the costs. These could include substantive changes such as output performance and over- or underruns on construction schedules or capital costs. They could also involve conventional changes such as fluctuating discount rates in the sector as a whole, lower discount rates tending to favour nuclear power. There could be fears about the economic risks of nuclear power in particular, leading to demand for differentially higher rates of return for nuclear projects.

The history of nuclear construction has not been a felicitous one. Many projects have suffered major overruns on projected construction times and costs, despite the initial optimism of its proponents. As mentioned, evidence from the Pacific Rim suggests that present construction programmes for proven designs are being carried out efficiently. Nevertheless, unless and until a significant programme involving an innovative design has been carried out successfully, it is likely that potential investors will still regard the construction phase as a risky one.

As nuclear power stations have relatively low variable costs as a proportion of total costs, one would expect that, once built, they would be used to supply baseload power. This economic tendency is reinforced by technical requirements, which make it relatively difficult to increase and decrease the output from nuclear stations on an hour-by-hour basis. Most existing nuclear plants are used for baseload rather than for load following, the notable exception being a part of the French nuclear industry.

Changing one factor, the discount rate, can have a profound effect on the relative economics of gas and nuclear energy as sources of electricity. There has been a vigorous debate about what discount rates are appropriate. Because high discount rates tend to value the present more than the future, it has been argued that lower discount rates should be used within a sustainable framework in order adequately to value the well-being of future generations of people. (This would of course require an entirely new approach to determining the economics of major projects.) Whatever the merits of these arguments, it is an undeniable fact that the discount rate being applied to power projects in most developed countries has been increasing as markets have been liberalized in recent years (see over), and it now often stands in the region of 12 per cent or more.

By contrast, the economics of gas-fired capacity tend to be relatively insensitive to factors such as problems during the construction phase or increased

Table 3.4 Gas price to US electricity utilities, 1997–2001

	1997	1998	1999	2000	2001
Gas price ($ per thousand ft³)	2.78	2.40	2.62	4.32	4.77
Fuel cost (¢ per kWh)	1.84	1.59	1.74	2.86	3.15

Source: USDOE (2002).

discount rates, but are far more sensitive to the costs of the fuel itself. At a time of rising fuel prices, this can be an important factor. Table 3.4 quotes the gas price to electricity utilities in the United States in recent years. It converts this into the fuel price component of the cost of gas-generated electricity, assuming 50 per cent thermal conversion efficiency – 1,000 cubic feet of gas will generate 150 kWh.

The investment and O&M costs of gas-fired electricity come to about 1 to 1.2¢ per kWh, the rest of the overall cost being accounted for by purchases of fuel. The effects of gas price changes are therefore very significant. (By contrast, of course, nuclear and renewable sources of electricity are relatively unaffected by fuel price changes.) Gas prices soon declined from their peak of early 2001, but there is evidence that several countries, including the United States, are reviewing their energy options in response.

The structure of the gas contract market will be of particular importance to the economics of gas-fired electricity. Long-term take-or-pay contracts are often quite different from short-term, spot market trading, owing to the different way in which economic risk is apportioned between gas producer–deliverer and power generator. As the variable costs are higher as a proportion of total costs, one might expect gas-fired capacity to be used to fulfil medium load and peak demand, all else being equal.

The effects of market liberalization

The command-and-control model

The changes in the organization of energy supply systems in many developed countries over the past decade or so have been profound. Historically, electricity supply systems were characterized by heavily regulated monopolies. In the United Kingdom, for example, the state-owned Central Electricity Generating Board, which also owned the grid, was in effect the monopoly generator of electricity in England and Wales. Area boards held local monopolies in electricity supply.

In the post-war years and up to the early 1980s, this command-and-control model was prevalent in most developed countries, although in some cases the utilities were privately owned rather than state owned. It persisted through the period of general concern, especially in Europe and Japan, that hydrocarbon fuel supplies (basically oil at that stage) were subject to interruption and likely to exhibit sustained high prices for the foreseeable future. This perception was of course fuelled by the enormous price rises and the fear of a reduction in the availability of oil in 1973 and in 1978–80, which had disastrous effects on the world economy in the middle to late 1970s and beyond.

During this time, governments tended to regard energy (or at least electricity and gas) as an industrial, or even a social, service: the role of government was to ensure security of supply. By regulation, and sometimes also by ownership, governments granted a geographical monopoly at some level in the system – either generation or supply or both. The generators and the local supply companies therefore had a secure customer base to which they could pass on any costs incurred in generating or procuring electricity as long as they satisfied the relevant regulator. Also, the generators could invest in sources of energy that would be unattractive in a more competitive market. In principle, diversity of supply could be imposed on the system to a greater extent than would occur in a free market, and thus security of supply could be safeguarded against overdependence on any one fuel, especially an imported one. In return, the utility with the monopoly accepted a 'duty to supply' electricity to anyone who wished to buy it now or in the future. In this way the degree of overcapacity and the combination of fuel sources necessary to ensure security of supply, both on an hour-by-hour basis and in the long term, could be safeguarded, at least in principle.

This is not to say that economic considerations were ignored entirely. The importance of reasonably low electricity prices to a country's international competitiveness was recognized. In the middle to late 1970s in particular, it was widely assumed that nuclear power would be the most economic source of electricity over the lifetime of the stations then planned. However, investors were aware that even if this assumption were to prove wrong, the economic downside would be limited. Provided the regulator could be satisfied, any excess costs of investment could be passed on to the customer in future electricity prices. It is fair to say that security of supply was the overriding consideration.

In these centralized systems, utilities were, to a considerable degree, instruments of state policy. The state might, as a case in point, wish to subsidize industrial users of electricity, either by offering a subsidy using taxpayers' money or by charging higher tariffs to domestic users. Even when privately owned, companies would accept a utility return on their capital (typically about five per cent), rather than the level of return they would demand in a competitive market (often as high as 15 per cent), in view of the low levels of risk involved. Their activities would be heavily influenced by signals from government.

Electricity generation tended to be carried out in large centralized plants, often of several thousand MW capacity, to benefit from economies of scale in production. Expenditure on long-term research and development, both directly by government and by utilities on government's behalf, tended to be sizeable, in keeping with government's perception of energy as a service to be guaranteed by the state.

The competitive model

After 1980, perceptions of high prices and unreliable supplies in world hydrocarbon fuels diminished. The oil price fell heavily, from $70 per barrel in 1980 to $20 in 1986 (all in 1998 money). For much of the late 1990s it stood at between $10 and $20 per barrel, quite typical of the period from 1930 to 1972, before the first oil shock of the 1970s. The amount of reserves of hydrocarbon fuels also changed dramatically. Proven reserves of oil in 1980 were 660 billion barrels; in 2000 they stood at 1,046 billion barrels (142 billion tonnes). Discoveries of gas were even greater: reserves doubled from 75 trillion cubic metres in 1980 to 150 trillion cubic metres (135 billion tonnes of oil equivalent) in 2000.[5]

In these circumstances, perceptions that hydrocarbon fuels were likely to be disrupted or to face sustained increases in price faded rapidly. At the same time, many developed countries had significant levels of overcapacity in generating plants of all descriptions. The priority afforded to ensuring secure energy supplies therefore declined, and with it the case for interfering with markets.

From the mid-1980s, then, the view of many governments in the developed world has increasingly been that energy is simply a traded commodity, the supply and price of which should be determined by the market. Market forces should also determine the level of diversity of supply. In this model, government's role should on the face of it be limited to ensuring fair competition (or regulating prices in those areas where there was a natural monopoly, for example distribution via national–regional grids and local networks) and

---

[5] BP Amoco (2001).

possibly environmental protection, as well as fulfilling the usual regulatory functions concerning health and safety etc. Since the 1980s, the main trend has been towards privatization and liberalization of energy supply systems, although at different rates in different countries.

The term 'liberalization' can cover a variety of different ways of organizing electricity market structures, each of which will have considerable regulatory input of various types from governments and government-appointed regulators. Nor does the existence of a more or less 'free' market in electricity prevent distortion of the market for other aims, for example to encourage the development of renewable energy sources. In some cases this might be done by simple operating or capital subsidy, in others by creating guaranteed markets for their output. In most cases the stated aim is for subsidies or market guarantees to be a short-term measure for aiding relatively new technologies to reach technological and economic maturity.

However, the structure chosen for the eventual market could also have profound implications for the relative prospects of different fuel sources. Examples of important issues in market structure include:

- the treatment of intermittent power output from, say, wind generators, especially if the proportion of the market supplied by that source should exceed about 20 per cent;
- the scope within the market for bankable long-term contracts, which may be of importance to the financing of longer-term energy projects. (It can be argued that even in those cases where long-term contracts have been allowed, those contracts are often broken if circumstances change, either by renegotiation or by the bankruptcy of one party. It is unlikely, in any case, that a competitive market would sustain the very long-term contracts necessary to underpin traditional 1-GW-plus nuclear projects with long construction periods.)
- the framework for financing back-end nuclear liabilities, mainly waste management and decommissioning, for example whether there will be requirements to establish segregated back-end funds and at what level of financing.

The costs per unit of output of electricity production have fallen markedly over the past 10 years. Generating companies have become much more cost-conscious in response to competitive pressures. The entry of natural gas as a fuel for electricity production, as a result of increased availability and the development of the combined cycle gas turbine, has also reduced overall costs.

However, companies in competition require higher rates of return than those operating as monopolies because they face higher economic risks. In particular, there is no longer a guaranteed market for electrical output at prices that would ensure long-term profitability.

For nuclear power, this is particularly detrimental. The higher rates of return demanded in the competitive market damage the relative competitiveness of highly capital-intensive projects relative to those with lower capital costs (notably natural gas). In addition, projects that take longer to recoup their initial investment are more risky than those in which initial costs can be amortized more rapidly. As a result, investors are likely to require a higher rate of return from nuclear projects than from those powered by natural gas, thus further increasing the economic advantage enjoyed by gas-fired projects.

Usually accompanying liberalization has been a tendency for the average size of new investment to fall from the 1-GW-plus units characteristic of command-and-control systems to several hundred MW or (sometimes much) less. (Privately owned power plants tend to be smaller in any case, and the development of the CCGT has undoubtedly been an important factor here as well.) At the same time, some commentators detect the beginning of a trend towards more localized power production at the expense of national or regional grids. In this structure, small power plants, including perhaps (at the extreme) 'microturbines' based in individual households, would be embedded in local demand networks.

Should this trend continue, it might become increasingly difficult to manage large plants within the national or regional system. The individual companies within each country are, by the nature of the system, smaller than the monopoly suppliers they have replaced. Thus, prima facie, they might be expected to be less able to finance large investment projects. (Recently in liberalized markets there have emerged cross-boundary operators, such as Electricité de France, E.On (Germany), British Energy and TXU (United States), which have assets in many countries. These international companies may of course have very large asset bases, and thus may be able to fund the construction of large power stations mainly from cash rather than from borrowing.)

Individual issues that may be especially risky for nuclear power include:

- the long timescale for construction, during which external events, changes in government regulations etc. may necessitate mid-project redesign, delay or abandonment;
- the perceived risks of project cost overruns;
- the risk of government interference, owing to either unhelpful external events or political considerations, including in the extreme case the elec-

tion of a new government with a less favourable attitude to nuclear technology, for example in Germany;
- the absence, in some countries, of coherent, long-term government policy
and support;
- the perceived risk of technical problems reducing electrical output, especially in the case of innovative reactor concepts;
- the availability to non-nuclear competitors of energy investment alternatives with lower initial investment costs and therefore lower risks in times
of stable fuel prices, notably natural gas;
- the need to cover engineering and licensing costs for the first of any innovative reactor design.

As noted earlier, the record of major construction projects has improved in
recent years, and evidence from the Pacific Rim suggests that nuclear construction costs are under control. Although this improvement would benefit
the economics of all power generation projects, the benefit in the case of
nuclear energy might be expected to be proportionally greater. Overruns in
construction costs or schedules are more serious for heavily capital-intensive
schemes. Controlling these costs would therefore disproportionately reduce
the risks associated with nuclear projects, and thus might be expected to reduce the differential between demanded rates of return for nuclear and
non-nuclear projects.

The perceived stance of government is important. In some countries, such
as Germany and Sweden, there is a history of governmental or public opposition to the nuclear industry. In others, such as Japan, France and Finland, the
national government is perceived as being less obstructive and/or public opinion appears to be more favourable. One would expect a potential investor in
new nuclear plants to demand a far higher rate of return on a project situated
in the former countries than in the latter.

However, liberalization of electricity markets has also had beneficial effects
on nuclear technology. In the sphere of operating costs, the remarkable improvements in the availability of nuclear plants in the United Kingdom and the
United States in recent years, referred to earlier, were a more or less direct
result of the commercial pressures of more competitive markets. Although
the pressure of liberalization on operating costs has been similar whether the
technology is nuclear or non-nuclear, the impact has been greater on nuclear
power because in many instances plants were being operated much further
from the achievable efficiency frontier in the nuclear than in the non-nuclear
cases.

A second area where liberalization may differentially benefit nuclear power is in construction costs. Nuclear procurement in many countries has tended to be nationalistic. The local utility usually has had substantial power over the chosen design, which has often been customized to a very high degree, and it has often been the project manager. Local firms have been preferred over international competition. The result has often been expensive.

In the liberalized climate, it is more likely that the utility, under shareholder pressure, will seek to buy a 'turnkey' project, or at least to buy major components of the plant, on an internationally competitive basis. The prospect, perhaps analogous to the aircraft industry, arises of a relatively small number of competing international designs, each licensed to operate in a number of countries. This should in principle lead to major cost reductions. Again, although the same forces will act on competitors to nuclear power, they are likely to have less impact. This is because procurement and management practice in CCGTs has approximated to the 'improved contracting' model from the start and has less room to improve and also because nuclear investment costs, being much higher, can in principle fall further. The nuclear construction market has become more international in recent years; and consolidation has led to the formation of international groups such as BNFL–Westinghouse–ABB and Framatome–Siemens.

A related factor is the growth of cross-border power markets. At present, the nuclear power stations in France must of necessity change their output as demand for electricity rises and falls through the day and through the year, as the amount of installed nuclear capacity is greater than the requirements for baseload in that country. More cross-border interconnectors, such as those that already exist between France and countries such as Belgium, Germany, Spain and the United Kingdom and allow France to export over 10 per cent of its electrical output, would enable those plants to operate at full capacity. There would be significant reductions in average unit operating costs.

Further, it is wrong to assume that there will be no players in liberalized markets willing to take a longer-term view. Institutions such as pension funds, for example, will be more interested in longer-term growth than in short-term returns. However, they will still be influenced by perceptions of political and commercial risk. For nuclear power to appear attractive, long-term costs and incomes will have to look favourable; and there will have to be confidence that onerous new regulations, or direct government interference, will not damage the project in the longer term. (There are, of course, risks associated with fossil fuel-fired plants as well. The increase in the oil price in the mid-1970s resulted in the abandonment of a large number of oil-fired power stations.

This was because their operating costs rose considerably and growth in electricity demand slackened in the recession.)

A further issue associated with liberalization is the nature of the contract market that might develop. (The weaknesses of the arrangements in California are considered on pages 83–84.) In the United Kingdom, recent changes to the market structure is putting a high degree of emphasis on reliability of supply, to the detriment of intermittent renewable sources, such as windpower. Companies relying on such sources of energy risk having to pay high prices for power in a 'balancing market' in order to fulfil their contract obligations at times when their wind generators are not generating. (The issue demonstrates the importance of considering the 'system costs' of producing reliable electricity supplies as well as the costs of individual plants.) Other models of the market place will doubtless develop, but the balance among cost, system security and ensuring market opportunities for intermittent power sources will be difficult to strike.

The future of liberalization

Should liberalization continue to spread through electricity markets in developed countries, it is likely that the average size of a plant will reduce. Financing large plants as existing units are decommissioned may become more difficult. These plants would have no guaranteed market for their output and would present high economic risk unless long-term power contracts can be signed. Experience of liberalized markets suggests that contract periods are quite short, at least where there is a surplus of generating capacity. In addition, increased support for renewable sources of electricity will increase the number of small units.

If this trend continues, questions will arise about the ease with which large plants of any description can be integrated into the grid. When 1,200 MW of electrical output goes off line, say for routine maintenance, it will be difficult to replace it with a large number of very much smaller plants. Similarly, to bring a large plant on line will require the retiring of plants owned by a large number of small utilities. This is likely to be a more difficult operation than replacing one large plant with another, and may create a market for smaller generating plants such as the 110 MW South African Pebble Bed Modular Reactor (PBMR), discussed later. It is perfectly possible that large plants would still be attractive for providing baseload power. On the other hand, the growth of localized grids with embedded generation units might, in the long term, reduce the baseload as a proportion of the total market.

The rapidity with which perceptions can change should again be noted. In the mid-1990s, for example, the accepted wisdom was that the vast discoveries of hydrocarbon fuel reserves over the previous two or three decades made a sustained repeat of the oil shocks of the 1970s, or of a shock to gas prices, look extremely unlikely. It was assumed that new electricity capacity, which the free market would build, would be predominantly gas-fired, at least where gas was fairly easily available. However, the tripling of oil prices in 1999–2000 and simultaneous increases in world gas and coal prices, along with questions over the future availability of gas in some areas, notably the United States, have qualified those certainties. It is even possible to envisage growing uncertainty over the very concept of liberalization of electricity markets.

Security of power supplies in liberalized markets

The introduction of competition in electricity supply systems has been accompanied by the disappearance of a 'duty to supply'. As no individual generator or supplier within the market can be sure of a future market, it is not possible to impose on any individual player a duty to supply in the future. While international fuel supplies are secure and there is ample generating capacity, it may not appear to be a significant weakness that energy is no longer to be regarded as a 'social service'. Also, the growth of cross-border electricity supply and trading offers the prospect of more secure supplies than might be the case in systems isolated by state borders.

However, electricity differs from most commodities in that it cannot be stockpiled. The difference between the highest and lowest demand during the year can typically be a factor of four or more. To ensure secure supplies at times of peak demand or when several large plants are off line for maintenance, it is necessary to keep a considerable margin of capacity, including some plants that may operate for only a small fraction of the year.

Security of supply presents few difficulties, in principle, within the command-and-control model of power production systems. Because the monopoly utilities have a duty to supply all consumers, they can invest in sufficient plant to ensure that they can do this. Any excess costs will be passed on to consumers in one way or another, as long as the relevant regulator agrees.

Reliability of supply is more problematic in the competitive market place. Presumably, the initial effects of capacity shortages would be power cuts at times of highest demand. To prevent this, new capacity would have to be ordered at least two years beforehand.

However, anyone investing in new capacity at that point in the cycle would be doing so for only one of two reasons: to compete with existing baseload plant, which would obviously have sunk its capital costs and so be able to operate at a lower marginal cost than new entrants, or to compete for the peak market, in which case the investor's plant would be operating only for a small amount of the time. In either case, the prospects for acceptable rates of return in the early years of the plant's operation might be limited. Although the long-term implications of this issue cannot be predicted yet, it is receiving increasing attention.

As time passes and more plants are decommissioned, new investment will become increasingly attractive. One can envisage a cycle of under- and overcapacity, with wide swings in prices similar to those observed in other very capital-intensive areas, such as the chemical industry. As operating capacity margins decline owing to retirement of older plants, market prices for the commodity tend to rise. These higher prices stimulate new construction, and a 'dash for investment' often materializes as companies fear being excluded from an apparently lucrative market. After perhaps two years, in which the companies involved make profits, there is often a further period of overcapacity, and prices fall towards marginal costs. This might be followed by a period of consolidation and low investment until capacity margins fall and prices start to rise again.

This pattern may be more severe for electricity, a product which, as noted, cannot be stockpiled and for which the difference in demand between peaks and troughs is so great. By the time the new wave of investment has led to the commissioning of new capacity, it is possible that there will have been prolonged power cuts at peak demand. Furthermore, it is at least possible that after the first such cycle, companies that had invested heavily but had achieved prices only at or about marginal costs might be more cautious about future investment. In any case, whether any of the investors in new power capacity are likely to make a sustained profit is questionable.

The experience in California in 2000–1 raised questions about the security of electricity supply in a (nominally) competitive market with capacity shortages. Recent studies suggest that the particular way in which the Californian market was organized and regulated was a major factor behind the problems. However, it is also possible that the problem lies at least in part in the concept of liberalization itself. Much analysis will be necessary before a firm view can be reached. A further aspect that demands attention is the short-lived nature of the crisis: by summer 2001 wholesale power prices were falling in California, owing in part to a particularly mild summer and the consequent reduction in power demand but also to energy-saving measures, falling gas prices and the slowdown of the US economy. Perhaps the most important lesson is that relatively small variations in demand can wreak havoc with electricity prices, at least when demand and capacity are finely balanced. This experience, suggesting that the period of high prices and high profits can be very short and unpredictable, might make investment in new power capacity even more unattractive.

In an oversupplied market, nuclear plants, if any were built, would occupy the baseload by virtue of their lower marginal costs. A reversal of market liberalization could of course lead back to a longer-term attitude to power investment – indeed, that would be the purpose. Although this change would not of necessity lead to a resurgence in investment in large nuclear plants, it would remove one of the obstacles – that of raising the large amounts of capital needed for such projects in the absence of long-term contracts for electrical output. Should liberalization continue, it seems unlikely that nuclear plants with the high capital costs typical of those now in use will be attractive to investors unless high fossil fuel prices prevail for a significant period of time.

The role of government in a liberalized market

However far liberalization may proceed, there can never be a 'free' market in electricity. Governments will still wish to ensure that power supply systems meet requirements in areas such as the environment and heath and safety. At times of supply shortage for whatever reason, governments are likely to wish to take steps to ensure reliable production. The form these steps take in a particular country at a particular time will be a matter of 'energy policy', a term used in a variety of contexts, from significant intervention in the market to attempts to 'clear the way' for private investment.

Insofar as one aim of liberalizing the markets is to move the locus of decision-making about energy sources, levels of diversity, security etc. from government to the private sector, governments would be expected to play less of a role in determining the structure and combination of fuels for electricity supply systems than in the command-and-control model.

Some fundamental responsibilities will certainly remain with government. They are broadly the same ones that government might play with respect to any industry. They might include:

•   ensuring as far as possible that an effective market is established, including competition in electricity generation and end-user choice, for example reducing barriers to market entry;

- monitoring the market structure and possible anti-competitive behaviour by market participants in order to sustain effective entry into the market and effective consumer choice;
- establishing well-funded, well-staffed and competent regulators and anti-trust enforcement agencies to prevent the emergence of barriers to new entrants to electricity generation;
- creating a favourable investment climate by providing a credible and clear regulatory framework and streamlining the administrative process for planning consent as far as possible without jeopardizing other important regulatory concerns;
- ensuring an adequate supply of suitably qualified personnel, through provision of higher education and vocational courses;
- protecting health and safety;
- developing a framework for limiting local, regional and global environmental damage associated with energy production and use.

Even within these categories there are areas of dispute. For example, the extent to which governments in the developed world believe action should be taken against the threat of climate change varies considerably.

For nuclear power, governments have also accepted a liability in the event of serious accidents. The operating companies accept strict liability for the consequences of an accident up to a certain figure, at present about $200 million, and they must have insurance cover for this sum. It is recognized that there is insufficient capacity in the commercial insurance market to accommodate very high-consequence, very low-probability risks such as an accident on the scale of Chernobyl or a large-scale terrorist attack. By international treaty, the costs of a very large nuclear accident will be covered by governments. It can be argued that the same would apply in practice to very large accidents in other energy fields (for instance large hydroelectric dams) and in other industries (for example chemical plants). However, as is discussed further in Chapter 5, on nuclear safety, the fact that radioactivity is detectable at very low concentrations, unlike certain chemicals, makes it likely that remedial action after a major incident will be proportionally more expensive. In any case, it is a legitimate question why nuclear power should receive this support when other power sources are not offered it and how the implied costs should be treated.

On the other hand, a considerable body of opinion holds that the role of government, even in liberalized markets, should go rather further. In many countries, the ideology of liberalization has included an assumption that energy is to be treated as a normal commodity rather than as a social service and therefore that all decisions on system reliability and investment in general, including issues such as security of supply, should be left to the investor and the consumer. But many commentators argue that given the central importance of energy, and especially electricity, within a modern economy, it is unrealistic to pretend that there is no national dimension to power production. Experience over the past 30 years suggests that developed economies cannot adjust rapidly to energy shortages, such as of electricity or fuels for transport. Major failures in electricity supply systems, for example those observed in California in 2000–1, would have a severe effect on a country's economy and thus inevitably carry political consequences for the government of the day. Pressures are likely to increase as excess capacity falls, either because of plants closing at the end of their economic lifetime or because of their forced retirement for other reasons, for example fossil-fired power stations as a result of concerns about climate change.

It is argued further that the role of government should go far beyond the monitoring of system reliability, investment in generation, input fuel diversity and the extent to which security of supply is being ensured by the market. Ultimately, governments need to be prepared to intervene in 'free' markets in order to protect system reliability, the security of supply and 'affordable' power prices.

If further analysis of the Californian power shortages supports this view, careful consideration would have to be given to what forms intervention might take. At the extreme, governments might take steps to recreate elements of a command-and-control system, in effect sacrificing a degree of economic efficiency for improved security of supply. It should be noted that in some models of liberalization as they have evolved, governments seem at present to have access to few real mechanisms for securing these goals.

The actions of governments in crises are perhaps even more difficult to model than future energy demand or fuel prices. What is certain is that different governments will act in different ways, depending on particular national circumstances, political ideology, national traditions etc.

Nuclear power in developing countries

It might be expected that as the developing world's consumption levels rise towards those of the developed world, the range of issues relevant to both will increasingly coincide. The growing importance of developing countries in world electricity consumption is difficult to exaggerate. This is shown in striking projections made by the International Energy Agency and others of patterns of electricity consumption (see Table 3.5).

Table 3.5 Projected electricity consumption, 2000, 2020 and 2050

Consumer	2000 TWh	%	2020 TWh	%	2050* TWh	%
OECD and economies in transition	10,500	70	14,500	56	14,600	41
Developing countries	4,500	30	11,500	44	21,100	59
Total	15,000	100	26,000	100	35,700	100

* Eden (1993).

From today's perspective, meeting the growing demand for energy in developing countries will probably place considerable pressure on fossil fuel production: renewables and nuclear power are unlikely to expand sufficiently in the next 20 or 30 years to play a dominant role. Oil and gas prices would be expected eventually to rise in response to this pressure, although it is often observed that the price of oil today has little relation to the costs of its extraction.

One might expect the same factors that affect nuclear economics in the developed world to apply, in principle, in developing countries as well. For example, it is unlikely that nuclear power will be attractive to the smaller developing countries for the foreseeable future in view of the infrastructure required to support a nuclear industry.

It is of course a mistake to view the developing world as a monolith. Even within a single developing country one can often discover enormous variations between, say, the more affluent urban and industrial areas and the much poorer rural regions. Access to natural gas, coal and other fuel resources varies at least as widely in the developing world as in developed countries.

However, some broad differences between electricity markets in developed and developing countries can be described. First, the rate of growth in electricity demand tends to be higher in developing countries. New capacity is likely to find a relatively secure market, and thus the economic risks of investment may seem to be lower than in a more slowly expanding market. In these circumstances large plants, with their economics of scale, may remain attractive in the bigger developing countries. Although there are questions over the raising of large amounts of capital in many developing countries, the need for new capacity of some kind means a considerable capital outlay, whatever options are chosen. Also, the installation of regional or national grids is an enormous challenge to many developing countries.

Second, some of the major developing countries face difficulties over the availability of fossil fuels. For example, in both China and India the main coal

mining areas are vast distances from the centres of economic growth, necessi-
tating major investment in rail links. Estimates suggest that the 'levelized'
costs of transportation in the coal-fuel cycle in China are between 0.48¢ and
0.76¢ per kWh, while the corresponding costs in the nuclear-fuel cycle are
perhaps one-tenth of this. A recent survey analysing the costs of installing
power capacity in Guangdong province in China found that the capital costs
of building a coal plant and upgrading the transport infrastructure were, at
$1,320 per kW, not greatly cheaper than those, at $1,600 per kW, of a nuclear
station of proven technology. In China, natural gas from the gas fields in the
East China Sea is required for the demands of non-power sectors, and transporting gas through 4,000 km of pipeline from Xinjiang in the west would
result in costs comparable to importing liquefied natural gas. As a result, the
relative economics of nuclear power and fossil fuel alternatives may be more
favourable than in countries with more economic gas supplies. (This is not, of
course, a point relevant only in the developing economies; it is also valid in
Japan, for example.)

Third, although environmental legislation has tended to be less stringent in
some developing countries than in the developed world, this is changing. It was
regarded as an important issue for China in its successful bid to host the Olympic
Games in 2008. In China, the ratio of the capital costs of nuclear power, based
on imported technology, to coal (without flue gas desulphurization) has been
about 2.5 to 3.3. However, a combination of the greater use of domestically
produced nuclear units and the need to fit FGD to new coal stations has reduced this ratio to 1.3 or 1.4.[6]

Fourth, market liberalization has only just started, if at all, in most developing countries. A degree of competition is being introduced in some areas.
China, for example, is in the process of separating generating companies from
grid enterprises. Issues about gaining a return on investment in electricity generation plants with high capital costs in competitive markets are likely to arise
if liberalization becomes more widespread.

Fifth, some commentators perceive that the political risks of investment in
some developing countries can be greater than in the developed world.

Finally, there was much debate before, and during, the Conference of Parties to the Rio Convention in The Hague and Bonn in 2000 and 2001 about the
potential role of nuclear power within the Clean Development Mechanism
(CDM) of the Kyoto Protocol. As discussed in Chapter 7, the CDM will allow
companies from the developed world to invest in projects in developing coun-

---

[6] Chuang-Yin (2001).

tries that result in lower emissions of greenhouse gases and to receive in return carbon emission reduction (CER) credits. These could be used to offset carbon tax or other obligations domestically, or could be traded on. Should there be a futures market in CERs, projected emission reduction could be used to defray the capital cost of a nuclear project. Although estimates that the CDM could generate credits equivalent in value to a 40 per cent reduction in nuclear construction costs do not appear credible, a figure of 20 per cent is more possible.

The nuclear industry and several developing countries argued that nuclear power should be included within the CDM, while opponents of nuclear power and some developed countries opposed the suggestion. The inclusion of nuclear power would of course have improved the relative economics of nuclear power in developing countries against fossil fuel alternatives. The development of a futures market in CERs and other tradable emission permits might then have allowed potential investors in nuclear power to gain access to a further stream of capital. In the event, it was agreed in Bonn not to accept CERs generated by nuclear schemes, at least within the first commitment period of 2008–12.

It is noticeable that no developing country commenced nuclear construction programmes in the past 10 years. The first nuclear plant in any country will be very expensive indeed, requiring not only the on-site and fuel-cycle facilities but also the establishment of national management and regulatory authorities. The reluctance of international financing bodies, notably the World Bank, to lend on nuclear projects exacerbates this situation. For a state to take this step, it will probably require a clear demonstration of the desirability of a nuclear industry of reasonable size or of a need to build up a base of nuclear expertise for other reasons.

In some developing countries the requirement for fresh water may be as important as the growing demand for power. Nuclear power is a potential power source for desalinating seawater – indeed, the BN-350 reactor in Kazakhstan was used for that purpose. The economics of this process might be improved if uranium could be extracted from the residue.

Future nuclear options

At least until recently, power plant sizes have tended to increase, especially in centrally organized electricity systems. The main reason for this has been the economies of scale associated with large plants. In the case of nuclear power in particular, there are a number of cost factors that are only weakly correlated with plant size. For example, the costs of salaries and overheads are not directly proportional to plant size. Certain elements of the capital plant, such as fuel-handling facilities, are required, whatever the size of the plant, and there are economies of scale to be had from equipment such as larger steam generators. The '0.6 rule' implies that if plant capacity doubles, capital costs go up by $2^{0.6}$, or approximately 1.5 times. This rule is applied widely to industrial projects, although there is a limit beyond which the relationship breaks down.

Considerations of economies of scale are likely to remain relevant in the future, in principle making large plants attractive in relatively large markets with power grids, especially markets with relatively little liberalization and in which demand is growing rapidly. China is a case in point. The average installed electrical capacity per capita there at the end of 1999 was 0.237 kW, about 40 per cent of the world average and a small fraction of the 3.29 kW in the United States. Electricity consumption per capita was 979 kWh, representing about one-half of the global average and between 10 per cent and 17 per cent of OECD levels. It is calculated that installed capacity will have to grow by some 120 GW to 240 GW by 2050.[7] As large amounts of capital will have to be found, whatever strategy to meet this requirement is followed, large plants are likely to be attractive, at least in areas with the heaviest economic development. Which fuel is preferred for such plants will depend upon a variety of factors, of which general economics in comparison with competitors will be central. Thus there will probably be a continuous demand for large plants, and there is no reason in principle why nuclear plants cannot compete.

In at least some liberalized markets, however, large plants of all kinds are less fashionable than they once were, for reasons discussed earlier. As a result, interest has focused on medium-sized plants of several hundred MW output and on small designs ranging from about 100 MW down to a few MW output in the case of some Russian concepts.

Furthermore, as noted earlier, liberalized markets, given their shorter-term focus, are likely to place greater emphasis on plants that can be brought on line quickly in response to developing shortages of capacity. Nuclear units of 1-GW-plus capacity tend to have construction phases of five years or more, and thus are unsuited to meeting such demands. Reactors with shorter construction periods, of perhaps about two years, would in principle be more attractive. They would also generate a cash flow earlier in the lifetime of the project than would plants with longer construction times. However, it should be noted that the licensing and planning process for new reactor construction in many developed countries is longer than the construction phase itself. This point would still need to be addressed should reactors with shorter construction phases become available.

---

[7] Ibid.

Evolutionary light water reactors (LWRs) such as Westinghouse's AP600 and AP1000, System 80+ and the Russian VVER 640 are being developed specifically to address the disadvantages of present-generation plants, in particular the high initial capital costs. Construction costs are reduced through simplified design (greater reliance on passive safety systems), modular construction, shorter construction times etc.

Some economic factors move in the opposite direction from economies of scale. For example, smaller reactors have a larger surface-area-to-volume ratio. This would make dissipation of decay heat in an emergency easier to achieve compared to a large reactor, with concomitant cost savings in the provision of emergency cooling devices. It is easier to use passive safety systems in smaller reactors. (Passive safety relies on forces of nature such as gravity or the boiling of water, while engineered safety relies more on human and technical features such as valves. It is argued that there is less need for multiple redundancy in passive safety systems, with consequent cost savings, as safety does not rely on a particular piece of machinery operating as designed.)

There could also be series economies, which would compensate for the loss of economies of scale at plant level. A modular plant consisting of 10 110-MW units might not benefit from economies of scale that would apply in a single 1,100-MW unit. However, building 10 identical units, especially on the same site, would offer 'nth of kind' cost reductions (that is series economies of scale). In time it is likely that a system consisting of a large number of small modules would develop a larger database of plant performance, beneficial in dealing with generic operational issues associated with the design in question. It is generally assumed that series economies will be even more significant in many renewable technologies as they become more widely deployed. Whether these opposing cost pressures would result in smaller or larger plants being more economic is likely to vary, and to depend on the particulars of the designs in question and on factors specific to individual projects.

Recent Westinghouse estimates for the AP600 suggest that a total generating cost of approximately 3.7¢/kWh should be achievable for an 'nth of kind' plant at a commercial discount rate of 11 per cent before tax. This is partly a result of lower investment costs (a reduced capital cost per kW(e) and a shorter construction time), lower operating costs and a higher load factor that should be achievable with simplified maintenance requirements. Fuel costs are projected to be similar to current-generation plants. Similarly, AECL claims considerably lower costs for the Next Generation CANDU reactor, owing to further reliance on passive safety features, a reduction in requirements for heavy water, higher fuel burn-up and the possible use of spent LWR fuel for new fuel assemblies. However, these claims may be difficult to evaluate precisely until a demonstration plant has been constructed.

The so-called 'third generation' nuclear plants include more radical departures such as high-temperature gas-cooled reactors (HTGRs). Particular interest has been generated by the Pebble Bed Modular Reactor being developed by Eskom in South Africa in association with international interests such as BNFL Westinghouse. HTGR technology was developed some decades ago, and several pilot plants were constructed and operated with some technical success. However, the technology was abandoned because of lack of competitiveness with LWRs and because of the failure of the commercial prototype at Fort St Vrain in the United States. It is estimated that new developments in direct gas-turbine technology are sufficient to overcome the technical obstacles that would previously have prevented HTGRs from being competitive. At the same time, the technical requirements for new plants have changed – safety requirements are more stringent, for example – and, it is argued, HTGRs are well suited to meeting these requirements economically. Engineering design studies carried out by members of the PBMR project suggest that HTGRs will be competitive economically with CCGT plants, although of course the transition from design study phase to commercial exploitation is often a difficult one. The suitability of HTGRs will remain speculative until demonstration plants have been constructed. Projected construction time is of the order of two years.

Incidentally, one non-power area in which nuclear technology may have a role to play is in providing process energy for extracting usable fossil fuel fractions from unconventional sources of oil such as oil shale. In some cases the extraction process can use half the energy contained in the oil itself. It is suggested by some commentators that HTGRs may be especially appropriate for this purpose as they produce very high process heat temperatures.

If this and other concepts are successful in reducing the capital-cost-per-unit-installed-capacity down to or below those of 1-GW plants, they may be more attractive than single large units. The capital outlay on a single small unit would be significantly lower and therefore more likely to be within the scope of relatively small generating companies. On the other hand, it is not clear that a generating company would wish to operate one or two small nuclear units (given the associated overheads etc.) unless there were special circumstances.

Furthermore, there may be benefits, at least in theory, in modular construction. Building a number of units on a rolling basis, all else being equal, may be expected to:

- allow a phased use of construction equipment;
- offer the prospect of a cash flow earlier in the project's lifetime;
- allow the spread across a number of units of the risk of overruns on the costs or timetable during construction or of poor output performance;
- allow rapid construction of the early units, in order to respond to an impending crisis in electricity supply.

Some commentators argue, however, that in practice, optimum plant size, at least for light water reactors, will remain in the region of about 1 GW or even higher, depending on market conditions. Alongside work on smaller reactors, parallel research is being carried out into concepts such as the American AP1000 and the European EPR-1750.

It seems unlikely that very small electricity markets, below, say, 1 GW in size, could, or would want to, develop the necessary infrastructure to support a nuclear power industry. But it is contended that areas such as Russia's northern and northeastern regions could be an exception. Some 10 million people live in those areas. They are engaged mainly in mining raw materials such as gold, nickel, tin, lead and tungsten. The profitability of these minerals' extraction, and thus the local population's livelihood, depends on reliable and economic sources of energy. The fall of communism has revealed the unsustainable economics of traditional approaches to providing energy to these industries, given the difficulties of supplying oil and other fuels to such remote territories.

One way forward, it is suggested, might be the provision of transportable nuclear units, based on reactors used for icebreakers and submarines, of a capacity from 1 MW to 50 MW. These units might operate for some 10 years without requiring refuelling, and would then be returned to the fuel processing plant and replaced by a new unit. A particular unit could be moved several times in its operational lifetime, depending on the requirement for energy. The claimed advantages for transportable units include:

- the use of autonomous energy sources very close to the source of demand, thus avoiding long-distance high-voltage transmission lines;
- the possibility of producing heat and energy for desalination of sea water;
- ease of transportation of the units;
- the customer is not involved in any operations with nuclear fuel management.

It will, however, be some time before this technology can be demonstrated on a commercial scale.

Already, however, there is a growing amount of interest and initial research and development work relating to revolutionary, so-called 'fourth generation' plants. It is claimed that these have the potential for significantly improved economics, including much shorter construction phases, simplified design and operation and improvements in safety performance. This issue is considered in more detail in Chapter 6, on research, development and commercialization.

It looks likely that should nuclear power once more become attractive, there will continue to be a major role for large reactors, but with rather lower investment costs than is now the case. The role for smaller reactors, either as modules in large plants or as stand-alone units, is inevitably more speculative.

The effects of market instruments on combating climate change

The command-and-control model of power markets offers a relatively straight-forward way of reducing harmful emissions to the environment. Given sufficient political will, utilities can be required to meet emission reduction targets. Although economic efficiency cannot be guaranteed, as it can be difficult to include cost signals in this model, in principle, environmental improvement can be assured.

In a competitive market, the environment can be protected by employing market instruments to punish companies for making harmful emissions. More-over, these instruments tend to apply a constant pressure on operators to reduce emissions, in contrast with the usual command-and-control tool that sets an emission limit but provides no incentive to improve performance beyond that limit.

In the words of the OECD, nuclear power operates under national and inter-national regulations that impose stringent limits on atmospheric discharges and liquid effluents, and its operators are committed to containing its waste and isolating it from the biosphere as long as it may be harmful for human health and the environment. According to this view, the corresponding costs, including waste management, waste disposal and plant decommissioning, have been internalised, and are borne by the generators of nuclear-generated elec-tricity. This applies also to liability in the event of a major accident, although the total liability assumed by the industry is capped, with governments carry-ing the residual risk.

Other commentators challenge the claim that nuclear power accounts fully for its waste disposal costs. They point to the large shortfalls between the provi-sions that have been made for nuclear waste management and decommissioning in many countries and the actual size of the liabilities as now calculated. These shortfalls, they argue, which will inevitably have to be met by the taxpayer, leave little ground for confidence that full provision will be made for such costs in the future, especially as many of the costs have not yet been quantified. In addition, the potential risks of a serious accident represent an external cost. Should an accident occur, they contend, much of the cost would have to be met by society at large, through the terms of international treaties that limit the liability of nuclear operating companies. As discussed in Chapter 5, on nuclear safety, it is difficult to compare low-probability, high-consequence risks with more 'everyday' higher-probability, lower-consequence risks, even when the latter risks are estimated to have similar cumulative consequences. It does seem clear, for example, that people are more concerned about the former risks, all else being equal, and it can be argued that this should be reflected in estimates of the external costs of nuclear generation.

Several studies have been undertaken on the health and environmental impacts of several fuel cycles used for electricity production, notably the most recent Externe report from the European Commission.[8] The findings of this work are by their nature uncertain, depending heavily, for example, on assumptions about the damage that will be associated with climate change. Nonetheless, their general finding is that under routine operation, nuclear energy and renewables have low impacts. The ExternE study suggests that the externalities associated with coal-fired or oil-fired electricity amount to approximately 5¢/kWh, while the corresponding figure for the natural gas cycle is about 1¢/kWh, in which the biggest inputs are the effects of climate change and the health effects of releases of particulates and other contaminants. Nuclear power's externalities average about 0.35¢/kWh, wind 0.15¢/kWh and photovoltaics 0.5¢/kWh.

Pearce, in reviewing a wide range of studies,[9] has found that the central estimate for damages resulting from greenhouse gas emissions is $25 per tonne of carbon for a scenario in which carbon dioxide mitigation is widely adopted. It is $70 per tonne of carbon (at 2000 prices) for a scenario in which the world continues to develop without widespread reductions in the emissions of greenhouse gases. These costs would equate to 0.3 to 0.9¢ per kWh for natural gas cycles and 0.7 to 2.0¢ for coal-based cycles.

The nuclear generation cycle does not release gases or particulates associated with acid rain or urban smog. Some carbon dioxide emissions result from the use of fossil fuels in extracting uranium, manufacturing fuel and

---

[8] European Commission (2001).
[9] Pearce (1995).

constructing facilities. But they are small relative to emissions coming from electricity generation with fossil fuels, and are comparable to those associated with renewables. A single 1-GW nuclear plant generating 7 TWh per year would be expected to offset the emission of some 1.9 million tonnes of carbon per annum if it displaced coal and 0.8 million tonnes of carbon if it displaced gas.

At present, fossil-fired electricity capacity does not internalize the costs of emissions of acidic or greenhouse gases. These costs will be met by the taxpayer, either now or in the future. It looks likely, however, that this situation will change at some point in the future, perhaps through the introduction of carbon taxes and/or tradable carbon emission permits.

BNFL–Westinghouse has stated that the AP600 design would be competitive with gas-generated electricity if carbon taxes were to be levied at a level equivalent to 0.7¢ per kWh, or $61 per tonne of carbon. The gap would be smaller, or perhaps even disappear, given the lower projected costs of the AP1000, the EPR or the PBMR. (As noted on page 92, these claims are disputed, and are difficult to evaluate in the absence of a demonstration project.) The Nuclear Energy Agency has noted that at a carbon tax of $100 per tonne of carbon (equating to approximately $12 per barrel of oil, nuclear power would become competitive with gas and coal (with a 10 per cent discount rate) in all but one of the countries providing information. The necessary figure would be rather lower in many cases. A study of options for Shandong province in eastern China suggests that a carbon tax of $37 per tonne of carbon would be sufficient to make nuclear power competitive with coal-fired capacity. Naturally, sustained oil and gas prices at higher levels would reduce or eliminate the level of carbon tax needed to make nuclear power competitive with gas and coal.

These levels of carbon tax are not unimaginable – indeed, they are already exceeded in countries such as Norway – but it is by no means clear whether or when they might be levied in other developed countries. Furthermore, although a carbon tax would improve the fuel costs of nuclear power and renewables relative to fossil fuel alternatives, it would not necessarily affect perceptions that highly capital-intensive projects carry higher economic risks.

The main alternative to carbon taxation is a system of tradable emission permits. It is generally held, at least in principle, that carbon taxes confer predictable costs on businesses but unpredictable consequences in terms of emission reductions and that tradable permits offer predictable general emission reductions but impose uncertain compliance costs on the polluter. The evidence from sulphur and other trading schemes in the United States is that the traded price per tonne of relevant pollutants may prove to be rather lower than originally expected. It seems that efficient and competitive trading tends to lead to relatively low prices.

As noted earlier, trading in a 'futures market' for tradable emissions credits could be used as a mechanism to provide a further capital funding stream for new nuclear construction. However, the above observations, taken together with the political difficulties of moving to a regime with high tax or permit prices, suggest that the market or political value of avoided greenhouse gas emissions may stay quite low for a long period. This in turn suggests that most of the economic distance between nuclear reactors and CCGTs will need to be made up by market-based developments (or of course by sustained high gas prices) and that it is highly unlikely that tax or permit regimes alone could transform the relative economics of nuclear power. Nonetheless, there may well be marginal cases of lifetime extension or new construction where some financial recognition of low greenhouse gas emissions may make the difference between viability and non-viability.

Security and diversity of energy supplies

Secure supplies of electricity are of great importance to many activities in a developed economy. Depending on the length of time of power interruptions, the cost of a lost unit to the consumer is suggested to be between 10 and 100 times as great as the cost of a delivered unit. But although security of supply is an important issue, some commentators suggest that appeal to it is sometimes the last resort of promoters of particular technologies who know their costs are uncompetitive.

The security of power supplies can be threatened in two ways. First, as discussed earlier, the capacity margin of plant available to serve the market in question can fall (because of rapidly growing demand or underinvestment in new plant or both), leading to an excess of demand over capacity at times of high demand. Second, supplies of the fuel can be interrupted (or become economically unsustainable).

Naturally, concerns about the security of fuel supply will vary. Countries with significant indigenous energy resources are less likely to be concerned than those that depend heavily on imports. Concerns about secure supplies are likely to be more prevalent at times of political upheaval and high world fuel prices, which often coincide, than in more stable international conditions. In some developed countries, recent events, such as the increase in the oil price in 1999 and 2000, the petrol crises in some European countries in September 2000, the Californian power cuts and concerns over international stability after the terrorist attacks of 11 September 2001, have resulted in a re-emergence of concerns about the long-term security of energy supplies, concerns that had lain largely dormant since the early 1980s.

Diversity of energy supplies is generally regarded as being of importance only insofar as it contributes to their security. In many countries, secure supplies have been safeguarded more by subsidizing a particular indigenous energy source, such as domestically mined coal, than by maintaining a wide diversity of fuels for power generation. Countries may pay quite high prices to improve the security of supply when they perceive a threat. Clear cases in point are France's nuclear commitment and Japan's current willingness to pay high prices for liquefied natural gas, nuclear power and imported coal. However, although the nuclear power programmes implemented by countries such as France and Japan illustrate how security and diversity of supply may be reflected in national energy policies, they do not provide a means to quantify the economic benefit that those countries ascribe to security and diversity of supply.

With the possible exception of Japan, then, countries do not seem to value diversity in its own right very highly. On the other hand, the recently published European Commission Green Paper *Towards a European Strategy for the Security of Energy Supply* states that 'the best guarantee of security of energy supply is to maintain a diversity of energy services and supplies'. The Green Paper makes the point that it would not be practicable to stockpile sufficient oil or gas for withstanding a long-term disruption in imported supplies, although it would be more feasible to maintain large stocks of nuclear fuels, since they can be easily stockpiled, represent only a small part of electricity generation costs and are produced mainly in OECD countries.

Whether governments will perceive a greater need to intervene in markets in order to ensure security of supply in the future is scenario-dependent. Should international instability grow and the level of trade decline, the traditional concerns could become stronger. In a more cooperative and cohesive world with freer and increasing trade, security (and thus presumably diversity) may recede from policy-makers' concerns. This seems already to be the case in countries that have plentiful indigenous resources. Whether this is a wise position to take in view of the rapidity with which the international political scene can change, for example because of terrorist actions such as in September 2001, is perhaps questionable.

Empirical research into the value of energy security has focused on the two oil crises and the period in the mid-1980s when oil prices collapsed. The results of this research remain inconclusive, as assessments of the economic effects of energy supply disruption differ enormously from country to country and study to study.

The situation in Russia merits mention. Russia's stated energy policy is to rely more heavily on nuclear power at home in order to be able to maximize the amount of gas exported for hard currency.

However, it is clear that a major nuclear incident, even in a foreign country, could have considerable effects on security of supply in a country with a major nuclear component. New construction projects might be abandoned or deferred, new operating regulations may reduce the availability of existing plants, and there may be temporary or even permanent closure of nuclear capacity. This might especially be the case if the incident were to occur in a generically similar plant. These effects were seen in the aftermath of the accidents at Chernobyl and, particularly, Three Mile Island. Although this effect may be seen with respect to other energy sources – for example, the Piper Alpha disaster in 1988 had a significant effect on offshore oil drilling – it is likely to be on a larger scale in the case of nuclear technology.

At times of apparent threat to secure electricity supplies, a considerable value may be put on ways of improving the security of energy supplies, but how great a value this would be is difficult to quantify. Generally, however, it would be unwise for promoters of nuclear power to assume that security and diversity will significantly 'rescue' the technology from uncompetitiveness. Indeed, under some circumstances the argument could be used against it.

Summary

Solving the economic problems that have affected the nuclear industry in recent years would not, of itself, make nuclear power more attractive. Many other issues, such as waste management, nuclear safety, weapons proliferation and public acceptability, would continue to be relevant. However, in the absence of great concern about the security of energy supplies or climate change, a clear prerequisite for a nuclear renaissance seems to be that the economics of nuclear power would have to improve beyond those of the currently deployed technology, which is based largely on designs from the 1960s and 1970s.

Factors such as climate change policy and fears about the security of energy supplies may be of marginal benefit to nuclear power, but by themselves are unlikely to make the crucial difference. For example, there seems to be little doubt that new regimes of pollution taxation will increasingly come into play and that they will help the competitiveness of nuclear power at least to a small extent. However, the views expressed by Hans Holger Rogner and Lucille Langlois of the International Atomic Energy Agency in a paper to a World Energy Council e-conference appear to be widely held:

It is true that nuclear power offers governments the opportunity to achieve a number of national policy goals, including energy supply security and environmental protection, particularly by reducing air pollution and greenhouse gas emissions. But these policy-related 'benefits' are vulnerable to policy change, and are insufficient by themselves to assure a nuclear future. Similarly, the further internalisation of externalities – to a large extent already imposed on nuclear power – is a policy decision and it is unclear when and to what extent such policies will be implemented. Those who pin their hopes for nuclear growth on externalities or on the Kyoto Protocol – and ignore reform and the need to innovate – will be doomed to disappointment. The nuclear industry has to bootstrap itself to economic competitiveness by way of accelerated technological development and innovation.

In particular, new nuclear products more suited to the competitive markets already established in many developed countries are required. If nuclear power is to be available, as a response to short-term power shortages for example, its construction phase will have to be much shorter than it has been over the past two or three decades, and serious efforts will also be required to reduce the time taken to license a new application. The investment costs per unit of electrical output will need to fall, and the stations will need to be demonstrably more reliable, especially during their early years, than has often been the case.

New designs that have evolved in recent years go some way towards answering these requirements. The construction record of new stations built in the Asia-Pacific region in recent years, for example, has been good. Plants with simplified design characteristics (to reduce initial costs and improve reliability), modular features (to reduce construction time) and other design improvements (for example higher fuel burn-up rates) are emerging, and three such designs have licences in the United States.

However, it will be difficult for potential investors to evaluate the claims being made on behalf of the more advanced new-generation reactor designs until demonstration plants have been built. There may well be a role for government in creating the conditions in which these plants are built, so that a proper appraisal can be made of nuclear power's economic potential.

References

Audus, H. and P. Freund (1997), 'The costs and benefits of mitigation: a full-fuel-cycle examination of technologies for reducing greenhouse gas emissions', *Energy Conversion Management*, 38 (suppl.), pp. S595–S600.

BP Amoco (2001), *Statistical Review of World Energy 2001*. London: BP.

Chuang-Yin, W. (2001), contribution to RIIA Nuclear Enquiry, China National Nuclear Corporation, P.O. Box 2102, Beijing 100822, *cywang@mx.cei.gov.cn*.

Eden, R. J. (1993), 'World Energy to 2050', *Energy Policy*, March, p. 231.

Elliott, D. (2001), contribution to RIIA Nuclear Enquiry, Open University: Milton Keynes, *D.A.Elliott@open.ac.uk*.

European Commission (2001), *ExternE – Externalities of Energy, http://externe.jrc.es/.*

NEI (1981, 1986, 1991, 1996, 2001), *Nuclear Engineering International*.

OECD–NEA (1998), *Projected Costs of Generating Electricity Update 1998*. Paris: OECD.

Pearce, D. (1995), *Blueprint 4 Capturing Global Environmental Value*. London: Earthscan.

USDOE (2002), *Selected National Average Natural Gas Prices, 1995–2001, http://www.eia.doe.gov/emeu/mer/txt/mer9-11.*

4 Radioactive waste management, reprocessing and proliferation

Introduction

Most radioactive waste now existing has arisen from the military nuclear programme, from early research and from the operation of commercial nuclear power stations. Waste arising from future nuclear operations will be much smaller in quantity per unit of electricity generated. Nonetheless, the management of existing wastes, and the question of whether more waste should be created, are controversial issues.

For present purposes the International Atomic Energy Agency's definition of radioactive waste seems to be appropriate:

Any material that contains or is contaminated by radionuclides or radioactivity levels greater than the exempted quantities established by the competent authorities and for which no use is foreseen.

Radioactive waste arises during nuclear fission – the process that produces the heat in nuclear power stations – as large atoms, usually uranium or plutonium, are split into smaller atoms, creating large amounts of heat. These smaller atoms, the 'fission products', are in general intensely radioactive. Spent nuclear fuel, that is fuel that has been used in a nuclear reactor, typically contains about three per cent fission products, alongside 96 per cent unused uranium and one per cent plutonium.

Uranium and plutonium themselves are radioactive, much less so than fission products but remaining radioactive for longer periods of time. Depending on a number of factors, such as their purity and the prospects for future nuclear reactors of various types, uranium and plutonium recovered from nuclear waste may be regarded either as potential fuels or as waste. Finally, some materials, such as trace elements in the steel used in reactor cores, can become radioactive as they absorb very small (sub-atomic) particles called neutrons.

The issue of waste management has major economic implications, especially when it comes to managing the legacy of waste that arose during early research into nuclear power and the wastes that have arisen from the early reactor systems. However, economics has not been at the centre of the nuclear waste debate in many countries. The issue of nuclear waste stands out in their wider nuclear debate as the topic in which the perceptions, or at least the public statements, of supporters and opponents are furthest apart. Many within the nuclear industry do not see the long-term management of nuclear waste, say by burial in deep underground repositories, as an 'insuperable', or even very difficult, 'technical' problem, assuming that 'proper investigation' of site geology is undertaken. But opponents of the industry consider that it is producing 'vast quantities' of 'deadly' waste and 'does not know what to do with it'.

As in all matters nuclear, there is a wide variation of attitudes and approaches in different countries. However, in many there is an often-stated view that until an acceptable method of dealing with radioactive waste is in place, the further development of nuclear technology (or at least nuclear power technology – it is rare that other sources of radioactive waste, such as nuclear medicine, are included) should be prevented. For example, in the United Kingdom the Royal Commission on Environmental Pollution has stated:

The intractable problem is to secure public agreement on the design and siting of a secure long-term repository. Considerations of inter-generational equity embedded in the concept of sustainable development demand the solution of the waste management problem, to the satisfaction of both the scientific community and the general public, before new nuclear power stations are constructed.[1]

Similarly, in its recent Green Paper on energy the European Commission stated that:

Nuclear energy cannot develop without a consensus that gives it a long enough period of stability, bearing in mind the economic and technological constraints of the industry. This will only be the case when the waste issue finds a satisfactory solution with maximum transparency.[2]

Supporters of nuclear power comment that this stricture seems to be applied only to nuclear power. There is no similar statement, for example, about emissions of greenhouse gases from fossil-fuelled power stations or about a variety of wastes from other industrial processes. Furthermore, they argue, radioactive waste differs from some categories of industrial waste in that it becomes less dangerous over time.

---

[1] Royal Commission on Environmental Pollution (2000).
[2] European Commission (2000).

However, the timescales involved can be extremely long, especially by the standards of human society. Plutonium-239, as a case in point, has a half-life of 24,600 years: after 24,600 years only half of the plutonium atoms will have decayed.

Whatever the merits of these arguments, it seems clear that the absence of an established waste management approach could act as a brake on any future nuclear development programme. This could occur in two ways. First, a shortage of short-term storage capacity for operational wastes from power stations could make their continued operation impossible. Although new plants can be built with sufficient capacity to store the arisings of spent fuel through the lifetime of the plant, some existing plants do not have this capacity. Second, the absence of a waste management approach could result in public and political pressure against new construction or perhaps even continued operation of existing plants.

Waste management, then, meshes together matters of physical science and public perceptions. Many of the latter topics are discussed in Chapter 2, on public perceptions and decision-making. This chapter will focus primarily on the technical issues; but at a number of points, matters of public and political perceptions will be covered where they are directly relevant.

Effective waste management requires perhaps three elements, in addition to the critical matter of public confidence:

- the need for all wastes to be identified, packaged and stored safely (relying on passive cooling and containment systems, if possible);
- the need to develop an infrastructure, including a decision-making infra-structure, that can implement a long-term management strategy;
- the need to ensure that sufficient money will be available to put the strat-egy into practice at the appropriate time.

This chapter will focus mainly on the issue of highly radioactive waste, that is waste containing significant quantities of fission products and thus emitting heat and high levels of radiation. The disposal of intermediate-level waste is also a matter of considerable dispute in some countries. Similar arguments are relevant to it.

Sources and types of radioactive waste

Radioactive waste is generated by a number of activities, which include:

- nuclear power generation both for electricity and propulsion (for example, nuclear submarines), with implications for the whole of the nuclear fuel cycle;

- decommissioning of nuclear facilities;
- accidental arisings of waste from incidents such as Chernobyl;
- the military defence programmes of a number of countries;
- the application of radioactivity in medicine and industry;
- releases of naturally occurring radionuclides owing to human activity.

Radioactive waste consists of a variety of materials having different physical and chemical properties and emitting different types of radiation. There are no international standard definitions of waste, in part because there are considerable differences between the types of waste which are arising or have arisen in different countries. A single approach would therefore be unlikely to fit all circumstances. However, a common classification system, used by the International Atomic Energy Agency, is the following:

- *Very low-level waste* (VLLW) is radioactive waste that, it is considered, can be safely disposed of with ordinary refuse. The IAEA calls this 'exempt waste'.
- *Low-level waste* (LLW) consists of litter and debris from routine operations and decommissioning. It is primarily material contaminated with low concentrations of beta or gamma emitters (see Glossary), but may include alpha-contaminated material. It does not usually require special handling, unless it is contaminated with alpha emitters. Most countries have well-established disposal facilities for low-level waste, and these wastes will not be considered in this chapter.
- *Intermediate- (medium-) level waste* (ILW) is waste containing higher concentrations of beta–gamma contamination and sometimes alpha emitters. There is little heat output from this category of waste, but it usually requires remote handling. It originates from routine power station maintenance operations, and includes the claddings that contain the fuel when it is in the reactor core. Such waste can be further classified as *short-lived* (usually meaning radionuclides with a half-life of less than 30 years) and *long-lived.* Some countries, notably the United States and Canada do not use this classification category. The IAEA has combined the classification of low- and intermediate-level wastes into LILW but has maintained the short- and long-lived subdivisions.
- *High-level waste* (HLW) may comprise either spent fuel (usually in the form of a ceramic) or the highly active material resulting from the first stage of fuel reprocessing, depending on whether the country reprocesses its spent fuel or not. (Some countries, such as the United Kingdom and France,

deem spent fuel as a resource, whereas Finland, the United States and Sweden regard it as a waste. The status of separated plutonium is also a matter of some dispute in many countries.) Immediately after reprocessing, the waste is in a liquid form known as *highly active liquid waste* (HALW); it may then be converted into a glass-like form known as *vitrified high-level waste* (VHLW). By far the largest part of its radioactivity derives from the 'fission products' within the waste, most of which have half-lives of less than 1,000 years. However, VHLW also contains long-lived elements known as 'actinides', of which uranium, neptunium, plutonium and americium are the most important, as well as some longer-lived fission products.

Some countries choose to categorize alpha-emitting waste separately. For example, the United States has a category 'transuranic waste' (TRU), broadly equivalent to ILW, for covering some of the wastes arising from research laboratories, fuel fabrication plants and reprocessing plants. In March 1999 a deep disposal facility – the first in the world for highly active material – known as WIPP (waste isolation pilot plant) began operation in New Mexico for that waste, which historically has arisen mainly from military operations.

The issue of categorization is an important one, as the definitions of different streams of waste tends to determine the management options pursued.

Quantities of waste

Uranium is a very concentrated energy source: one tonne of uranium used in a typical pressurized water reactor (PWR) produces the same electrical output as approximately 20,000 tonnes of coal. Thus, the volumes of waste produced in modern nuclear stations are modest by industrial standards, and do not include emissions of greenhouse gases, notably carbon dioxide and methane, which are implicated in climate change. (Of course, the use of fossil fuels at some points in the nuclear fuel cycle, such as the extraction of uranium and plant construction, does cause some greenhouse gas emissions.)

As a rough guide, a 1,000-MW(e) PWR will use 32 tonnes of fuel, containing 26 tonnes of uranium, per year and produce 7 TWh of electricity (assuming an 80 per cent load factor). Without reprocessing, some 32 tonnes of spent fuel will be produced (containing about 25 tonnes of heavy metals, mainly uranium, neptunium, plutonium and americium) for storage or disposal, along with approximately 300 m³ of low- and intermediate-level waste.[3]

---

[3] NEA (1993).

Table 4.1 Materials arising from reprocessing light water reactor fuel heavy metal

	VHLW (m³)	ILW (m³)	LLW (m³)	Reprocessed uranium (tonnes)	Separated plutonium (tonnes)
Per reactor per year	2	20	75	24.5	0.25

Source: RWMAC (2001).

If the spent fuel is to be reprocessed, it is first dissolved, usually in nitric acid. The chemical extraction process that follows produces the quantities of materials shown in Table 4.1. One tonne of heavy metal spent fuel is equivalent to 0.4 m³, or 0.9 to 1.5 m³ after conditioning for disposal. In this sense reprocessing can be said to reduce the volume of highly active material for disposal. However, as discussed later, reprocessing does not affect the amount of heat being generated. As a result, a repository to take reprocessed waste would not necessarily be much smaller than one taking the equivalent amount of spent fuel. Further, reprocessing gives rise to more than half of the intermediate-level waste being produced in a country such as the United Kingdom or France. In comparison, a coal-fired plant of the same size would release some seven million tonnes of carbon dioxide per year and smaller quantities of sulphur dioxide (perhaps 2,000 tonnes), nitrogen oxides (3,500 tonnes), ash etc.[4]

Each year, nuclear power generation facilities worldwide (including fuel-cycle plants etc. as well as power stations) produce about 200,000 m³ of low- and intermediate-level waste. By 1998 global nuclear activity had produced some 200,000 tonnes of spent fuel, representing over 20 years' arisings of waste at the then current rates of production of about 8,500 tonnes per year.[5]

Clearly, the amounts of waste arising from the future use of civil nuclear power will depend on three factors:

- the nuclear power generation technology itself – plants with higher fuel 'burn-up' or better thermal efficiency etc. would result in reduced waste production per unit of power output;
- the techniques used in the fuel cycle, such as methods of enrichment and use of reprocessing;
- the number of operating nuclear power reactors and their electrical capacities.

---

[4] European Commission (1997).
[5] Häfele (1998).

Proposals for final waste disposal

There are few firm proposals for repositories for highly radioactive materials. Those proposed for Yucca Mountain in the United States and for a site near the Olkiluoto power station in Finland are possibly the most advanced in conceptual terms, with projected operational dates of 2012 and 2020 respectively.[6] It is difficult, therefore, to address the question of how many such facilities would be required for different levels of future nuclear activity. However, as estimates are required for the expected costs of deep geological disposal, most countries with nuclear power facilities have produced 'reference designs' for repositories. Although final repositories, if built, may differ considerably from these plans, they might nonetheless allow a feel for the number of repositories that would be required under different scenarios.

The size of these reference designs varies considerably, depending on the amounts of waste arising in the country in question, assuming that each country would be responsible for its own waste. Some reference designs would provide for disposal of highly active materials equivalent to less than 1,000 TWh of electricity (Finland, the Netherlands and Switzerland) and others for materials equivalent to between 1,000 TWh and 10,000 TWh (Belgium, Germany, Spain, Sweden and the United Kingdom). The largest reference designs, in France and the United States, would provide disposal for highly active wastes arising from nuclear output of about 25,000 TWh.[7]

In the year 2000, global nuclear electricity production was about 2,400 TWh. Should nuclear power decline over the next decades, one would expect its operation to require no more than one repository for highly radioactive material in each country, with a few possible exceptions, notably the United States. A continuation of nuclear power at or just above today's levels would require the equivalent of one repository on the American or French scale every 10 years or so, even assuming that future nuclear designs will have higher fuel burn-up and thus be more efficient in terms of waste production than the reactors being used at present, many of which were designed in the 1970s and 1980s. (Projections for the AP600 reactor, for example, suggest that spent fuel arisings per MWh should be about 20 per cent lower than for existing large nuclear stations, although this remains to be demonstrated.) Of course, should each country develop its own repository, as is currently envisaged and enshrined in a number of laws and conventions, the total number of smaller facilities would be greater.

---

[6] House of Lords (2001).
[7] NEA (1993).

However, if use of nuclear power were to expand to, say, 10 times the current generation levels by 2050, representing perhaps 50 per cent of world electricity production at that time, this would result in arisings of highly active waste requiring the equivalent of approximately one major repository on the American or French scale per year, assuming no radical change in technology. (There would also be considerable arisings of intermediate-level wastes that would require management.) In these circumstances, questions about siting and public acceptability would probably become more acute, raising questions as to whether new approaches such as partition and transmutation (P&T) of more active wastes should be pursued. (Even if successful, P&T would not eliminate long-lived waste entirely, although it would reduce its volume substantially.)

Larger repositories, if technically feasible, may become more attractive. This is because the unit cost of a repository is not directly proportional to its volume, several of the costs being more or less fixed,[8] for example:

- the supporting scientific programme;
- the development of waste packaging specification and guidance;
- site selection, investigation and characterization processes;
- the preparation of the safety case;
- public inquiries;
- surface works, such as the fuel reception area;
- surface and underground infrastructure;
- final closure and decommissioning.

In addition to the highly active wastes discussed on pages 105–106, storage will be required for the non-alpha-bearing intermediate-level wastes that arise in the course of producing nuclear power, including, for instance, some decommissioning wastes. In the United Kingdom, Nirex has estimated that to move from an intermediate-level waste repository of 200,000 m^3 capacity to one of 400,000 m^3 capacity may result in an increase in cost from $8.3 billion to $10.1 billion, an increase of about 22 per cent, although this calculation is extremely speculative and depends on a large number of assumptions.[9]

The costs of disposal of nuclear wastes are very large in absolute terms, and estimates of the total costs have grown significantly in recent years. Thus, the US Department of Energy, in its life-cycle cost assessment for the high-level radioactive waste programme in 1995, estimated the costs of the programme

---

[8] BNFL (2001).
[9] Finch (2001).

up to final closure in perhaps 100 years at $37.9 billion. The Nevada State Agency for Nuclear Projects estimated the total costs at $56.2 billion. The latter study estimated that the maximum revenue to the Nuclear Waste Fund established under the Nuclear Waste Policy Act of 1982, under which nuclear electricity producers pay the US government $1 per MWh of electricity generated, assuming that all licensed stations operate to the end of their design lives, would be $29.3 billion (in year 2000 money).[10]

Reprocessing

As noted earlier, when spent fuel leaves a civil nuclear reactor, usually after about five years of power production, it typically consists of 96 per cent unused uranium, one per cent plutonium and other transuranic elements and three per cent 'fission products'. The 'fission products' are the smaller atoms produced by fission of the uranium, many of which are intensely radioactive and which form the basis of 'high-level waste' if separated from the spent fuel.

At present, there are two broad approaches to managing spent fuel. It can be dealt with in the form in which it leaves the reactor – the 'once-through' concept – or it can be 'reprocessed', separated into three streams – uranium for potential reuse in the short term or for storage; plutonium for storage, use in fast reactors or use in mixed oxide (MOx) fuel; and high-level waste. The liquid waste needs to be immobilized in a suitable matrix for eventual disposal – glass (forming vitrified high-level waste) and synroc (an advanced ceramic developed in Australia in the 1970s) are two examples of such a matrix. High-level waste and spent fuel require remote handling because of their intense radioactivity.

(The next-generation Canadian deuterium-uranium reactor (CANDU), under development in Canada, is expected to be able to run on spent fuel from light water reactors. It is currently estimated that the spent fuel from three light water reactors (LWR) would be sufficient to provide all the fuel for one CANDU reactor of the same output. Eventually, however, spent fuel from the CANDU reactors themselves would require management.)

To these approaches can be added a third, as yet unproven, possibility (which has some similarities to reprocessing), partition and transmutation. This allows for the destruction of most of the longer-lived fission products and actinides while separating out short-lived fission products for storage or

---

[10] Loux (1998).

disposal and uranium for storage or reuse. This will be considered in more detail on pages 132–134.

At present, only a relatively minor proportion of the world's spent fuel, perhaps 20 per cent, is reprocessed. This figure is likely to fall, at least in the short to medium term. Countries that reprocess at least some of their spent fuel, either in their own facilities or by using facilities in other countries (notably France and the United Kingdom), include Japan, Russia, France, the United Kingdom, Germany, Switzerland and Belgium, although Germany has announced its intention to stop transportation to reprocessing plants by mid-2005. Official policy in China and India also favours reprocessing. The two main reprocessing facilities outside the former Communist world are at Sellafield in Britain and Cap de la Hague in France, both of which accept foreign fuel for reprocessing, on the basis that contracts signed since 1976 contain a provision for all products to be returned to the country of origin. Japan is officially in the process of building its own reprocessing facility, at Rokkasho-Mura, for commissioning in 2005, although progress has been slow. Among those countries that adhere to a 'once-through' approach are the United States, Sweden, Canada, Finland and Spain.

Reprocessing had its origins in the need to separate plutonium from spent fuel from military reactors for the weapons programme. As an approach to managing civil nuclear spent fuel, it was initiated for economic and resource reasons, based on perceptions of future uranium supplies and demand. It has attractions and drawbacks, as discussed in the following sections.

Arguments for reprocessing

In its favour, reprocessing can in principle reduce the amount of fresh uranium that must be mined and processed, by allowing recycling of uranium from the spent fuel. However, the uranium that is won back by reprocessing is more radioactive than fresh enriched uranium. The decision whether or not to recycle the uranium from reprocessing therefore depends on the relative costs of procuring fresh uranium, as against the additional costs of utilizing the more radioactive reprocessed uranium.[11]

Paradoxically, should reprocessed uranium be declared a waste form, this might increase the likelihood of it being recycled. Any waste, by definition, has negative value, as it requires resources for storage or disposal. If recycled uranium is viewed in this way, then the economics of its use in reactor fuel,

---

[11] Uranium Institute (1996).

from the point of view of the fuel manufacturer, will include a credit for avoided waste management costs. Obviously, this credit would not be available if the uranium were regarded as a resource and the fuel manufacturer had to pay for it. (The same argument would apply to plutonium, and explains the apparently strange observation that a country such as France can regard plutonium as having zero value but still use it in MOx fuel.) Of course, declaring recycled uranium and plutonium as wastes would significantly damage the economics of reprocessing.

Reprocessing also separates out plutonium, which is a potential reactor fuel in 'fast' reactors (FRs) and high-temperature gas-cooled reactors (HTGRs). It can also be used in more conventional thermal reactors in the form of mixed oxide (MOx) fuel, a mixture of uranium and plutonium oxides, usually in a ceramic matrix. (In some cases, notably British Magnox reactors where spent fuel has been wet-stored after coming out of the reactor, reprocessing has always been considered necessary for technical reasons.)

In the period between the inception of nuclear power in the 1950s and the end of the 1970s, it was generally assumed, at least within the nuclear industry, that reprocessing would be an essential element in future nuclear development. Although uranium had been discovered in the eighteenth century, few uses had been found for it before the discovery of nuclear fission in the late 1930s. As a result, little attempt had been made to discover fresh deposits of the metal, and it was assumed to be rarer than metals such as palladium and platinum. At the same time, it was projected that nuclear power would grow very rapidly: in 1974 the IAEA forecast that there could be up to 4,450 reactors operating by the year 2000[12] (the actual figure was 438). Given these two premises, it seemed clear that uranium for use in thermal reactors would be a very limited resource.

Natural uranium consists of 0.7 per cent of the isotope uranium-235 and 99.3 per cent of the isotope uranium-238 (and tiny traces of uranium-234). Thermal reactors, which include the vast majority of nuclear reactors operating today, in effect operate only on the uranium-235, which is 'fissile' – it undergoes nuclear fission when struck with neutrons, and so can be used as a source of energy. Uranium-238 does not undergo fission; but in a nuclear reactor uranium-238 atoms can absorb neutrons, and eventually turn into plutonium.

If the plutonium is separated from the spent fuel, it can be used in a different type of reactor, a fast reactor. Separating plutonium from spent fuel and

---

[12] IAEA (1974).

using it in a fast reactor in principle increases the amount of energy that can be obtained from a certain mass of natural uranium by a factor of about 60 (once the energy needed to carry out the separation etc. has been accounted for). This is because fresh plutonium can be 'bred' in a jacket of uranium-238 around the core of the reactor, where it would gradually be converted into plutonium by neutrons lost from the surface of the core. (Alternatively, FRs can be designed to be net users of plutonium, by operating them without the uranium-238 jacket.)

In the event, neither premise predicting a shortage of thermal uranium by the end of the twentieth century proved to be correct. Uranium has proved to be a relatively plentiful element in the earth's crust, commoner than tin and zinc for example. Moreover, there is a vast quantity of uranium dissolved in the oceans. Its extraction is technically feasible, and may become economic at uranium prices considerably higher than those of today.

At the same time, the world's nuclear industry retained its position as proportionately the fastest growing of the major energy sources in the 1980s and 1990s, but it did not expand as rapidly as was assumed before the fall in world fossil fuel prices in the 1980s, the effects of liberalization of power markets in the 1990s (especially on heavily capital-intensive technologies) and the enormous discoveries of fossil fuel reserves in the past two decades. As a result, the pressure on thermal uranium reserves has eased considerably, and the case for reprocessing on resource and, especially, economic grounds has weakened. Separated plutonium, which was expected to be a valuable commodity, is, as already mentioned, accorded zero value by the French electricity utility EdF; it is regarded as an economic liability in some countries, but still has a nominal positive value in others. Nonetheless, reprocessing as a possible way of reducing dependence on imported uranium still has advocates in countries, such as Japan, that do not possess significant indigenous energy reserves. Some commentators suspect that the export of spent fuel for reprocessing has been politically expedient, removing it from home soil at least temporarily.

In addition, the technical and financial experience of the fast reactor has been very patchy. The only commercial-scale fast reactor, Superphénix, was finally closed down in 1998, having generated only 8 TWh in its 13 years of operation, although its predecessor Phénix has been more successful. Russia's 600-MW BN-600, at Beloyarsk, has been markedly more successful, but is now fuelled with highly enriched uranium. Although a number of countries, including Japan, India, China and Russia, remain officially committed to the fast reactor, there has been little progress in recent years. The 250-MW research FR at Monju in Japan, for example, has been closed since December

1995, although there are plans to reopen it. FR programmes in the United States, the United Kingdom, Germany and Belgium have all been cancelled. However, the 400-MW Fast Flux Test Facility at Hanford in the United States has been kept ready for possible recommissioning.

If the uranium and/or plutonium were recycled, reprocessing would reduce the requirement for mining fresh uranium, and therefore reduce emissions and radiation doses from mining and mine tailings. Reprocessing also reduces the volume of highly radioactive material for storage or disposal. Reprocessing does not affect the overall radioactivity of the material coming from the reactors, but it concentrates the highly radioactive materials (mainly fission products and some of the minor actinides) into a volume roughly one-fifth of that of the initial spent fuel. Also, the reprocessed waste would not contain significant amounts of plutonium or other proliferation materials, although some commentators argue that the separated plutonium may represent a bigger threat.

However, as reprocessing does not affect the amount of heat generated in the highly active materials, the packages will still require similar spacing within any final repository. The result would be that the size of the repository for the highly active material, and its associated cost, might need to be similar for both reprocessed and unreprocessed fuels. In addition, of course, reprocessing would create intermediate- and low-level wastes that would need to be managed. It would also produce uranium and plutonium, and these too might be declared as wastes and require managing.

Drawbacks of reprocessing

Reprocessing also has drawbacks. It is expensive, as it requires considerable quantities of materials and plant. Thus it seems highly questionable whether any major new reprocessing plants will be built in the developed world in the foreseeable future, with the possible exception of the partially completed Japanese plant mentioned earlier. Furthermore, the materials and plants themselves become radioactively contaminated, and must therefore be treated as wastes when the facilities are decommissioned. The quantities of these wastes can be substantial. At present, reprocessing is based on an aqueous chemical process whereby the spent fuel is dissolved in hot nitric acid and treated with solvents and pulses of compressed air to separate the uranium, plutonium and waste streams (the Purex process).

Reprocessing, using current technology, also creates highly active liquid waste as an intermediate before this waste is vitrified. Although this material is stored in robust facilities, the issue of major damage to these facilities caused accidentally or deliberately is important, and there is a range of views as to what the effects of damage might be, including claims it could result in releases greater than those at Chernobyl.[13]

In addition to the sludges and decommissioning waste that arise from reprocessing, the operation of the reprocessing plants at Sellafield in the United Kingdom and Cap de la Hague in France has historically been responsible for much higher levels of environmental discharge than have the operating power stations. In 1999, for example, Cap de la Hague released roughly 15,000 times as much radioactivity as the nearby Flamanville 1300-MW reactor.[14] These discharges fell significantly in the course of the last quarter of the twentieth century, and the nuclear industry would argue that they are now of little environmental consequence. As a case in point, emissions of cacsium-137 from Sellafield fell from 5,000 TBq per year in the mid-1970s to less than 10 TBq in 1999, while total alpha-emitter discharges fell from over 150 TBq per year to 0.1 TBq per year. Doses to the most exposed individuals as a result of the activity of Cap de la Hague are below 0.2 mSv per year, against typical natural dose levels of above 2 mSv per year.[15] However, discharges are still the subject of considerable international dispute, as demonstrated by the call from the majority of the OSPAR[16] countries bordering the North Sea to reduce emissions from these plants to zero by 2020 (although their final decision demanded 'close to zero environmental impact').

Other reprocessing techniques, which do not require the use of water and which may therefore result in much lower levels of emissions, have been investigated in laboratories. They include pyrochemical techniques, which may also be of use during partition and transmutation (see page 132). However, they have yet to be demonstrated on a commercial scale, and would still face questions over the management of the highly radioactive products.

Central reprocessing facilities entail movements of spent fuel from the power stations where it arises. Although transportation has in general been safe, it is a source of political and public disquiet in some countries. The same transportation issues would arise, of course, should central spent fuel stores be built in a particular country, but not if the material were to be stored at the site of origin.

---

13 WISE (2001).
14 Schneider et al. (2001).
15 Strupczewski (1999).
16 OSPAR is the Convention for the Protection of the Marine Environment of the North-East Atlantic. It consists of 15 members and the EU.

The other main concern raised by opponents of reprocessing is its role in separating out plutonium from spent fuel. They argue that separated plutonium has potential consequences for the proliferation of nuclear weapons, as anyone wishing to make such weapons would not themselves have to separate plutonium from highly radioactive spent fuel.

The actual and estimated arising of plutonium since the discovery of nuclear fission in the late 1930s is indicated in Table 4.2.

The total amount of plutonium being generated in nuclear reactors is about 50 tonnes per year. In 1995 the stockpile of separated plutonium in stores increased by 14 tonnes (22 tonnes being separated through reprocessing, of which eight tonnes were used in fabricating MOx fuel).

Most of the world's existing plutonium, then, is retained within spent nuclear fuel. In this form the plutonium is not attractive as a potential source of nuclear weapons, as the high radioactivity of the fission products in the spent fuel renders handling extremely hazardous. In addition, much of the spent fuel is in ceramic form, from which it is physically difficult to extract materials. However, significant quantities of civil plutonium have been and are being separated from spent fuel in reprocessing, with the result that the projected amount of separated plutonium in 2010 (stored or in MOx fuel) is much higher than the 1995 figure. The implications for nuclear weapons proliferation are considered later.

It should nonetheless be repeated that reprocessing as such neither creates nor destroys radioactive materials. The plutonium remaining in spent fuel may represent less of a proliferation risk in the short term, because the intense radioactivity of the short-lived fission products renders its use unmanageable

Table 4.2 World plutonium stocks, 1995 and 2010

Source and use of plutonium by sector	1995 (tonnes)	2010 (tonnes)*
In spent reactor fuel	800	1,400
Separated and stored plutonium	140	150
In thermal MOx fuel	20	550**
In fast reactor fuel cycle	30	**
Civil subtotal	990	2,100
Military subtotal	249	n/a
Total	1,239	

* Estimated.
** For both MOx and FR fuel cycle.

Source: IAEA (1997).

without very large resources. However, this radioactivity will fall significantly in the course of one to three centuries. After this time, spent fuel disposal repositories could in principle become attractive sources of plutonium – 'plutonium mines', as they have been termed by opponents of the industry. Unless some way of destroying plutonium is employed, the long-term implications of managing plutonium, separated or not, could be serious.

The options for reducing the plutonium stockpile include:

- using plutonium fuel in fast reactors or high-temperature reactors;
- using plutonium in an inert fuel matrix in thermal reactors;
- transmutation;
- using MOx fuel.

The use of MOx fuel will not necessarily result in a net reduction in total plutonium. While plutonium is being destroyed, some of the uranium oxide is converted into fresh plutonium. The relative rates at which these two processes occur, and therefore the effect on the total amount of plutonium present, depends on the relative proportions of plutonium and uranium oxides in the fuel. Of course, if the reactor is fuelled partly by MOx fuel and partly by conventional uranium fuel, plutonium will be building up in the conventional fuel even if, on balance, it is being destroyed in the MOx component.

Major spent fuel management options

The management of spent fuel, as in any other field in which there are several requirements of a process, is often a matter of trade-offs. Keeping spent nuclear fuel on the surface allows heat to decay and thus aids eventual repository design (assuming that is the approach to be followed). But at the same time, the material remains more vulnerable to accidental or deliberate damage. Reprocessing with a view to using the plutonium may represent a short-term proliferation risk. In the longer term, destroying the stockpile of plutonium, both that which has already been separated and that which remains in spent fuel, may reduce proliferation concerns (although this would depend on other factors such as the number of plutonium or spent fuel stores and the transportation requirements of using plutonium fuels.) Transmuting long-lived radioactive materials, if successful, would reduce the lifetime of their activity, at the expense of increasing their toxicity in the short term and the requirement to carry out additional processing. Delaying decisions may allow more time to reach a societal consensus on the way forward, but may cause operational difficulties or short-term costs in those countries where interim storage facilities are becoming full.

Leaving aside the issue of whether or not to undertake reprocessing, discussions on waste management strategies have been dominated by discussion of the relative merits of disposal and storage. The term 'disposal' has come to mean emplacement of waste in a final repository (usually assumed to be deep underground) without facilities for retrieving the waste at any point in the future. 'Storage' means packaging and emplacing the waste in a temporary store, where it can be monitored, with a view to retrieving, repackaging and/or placing the waste in a final repository in the future, should it be necessary. Stores could in principle be positioned on the surface, at a shallow depth or deep underground.

When the fuel is removed from the reactor, it must in all cases be stored for an initial period, up to five years, for the extremely intense heat to dissipate and radioactivity to decay. Even after this time, the waste would be producing significant quantities of heat and radiation. One light water reactor fuel assembly, typically containing about 500 kg of uranium, generates heat of about 17 kW one month after discharge from the reactor, 4 kW one year after discharge and 0.8 kW five years after discharge. The heat output falls by a factor of 50 or more in the first century.

To attempt to dispose of the waste deep underground while it is still generating a significant amount of heat would cause extra problems, as a way would have to found of venting the heat to prevent the risk, for example, of the fuel melting. The Yucca Mountain facility proposed for the United States is conceived as taking heat-producing spent fuel, but at the expense of more sparse packing of the canisters, so reducing the volume of waste that can be emplaced in the facility.

A number of technical options have been put forward for managing highly active wastes after the initial storage period, but all present technical challenges. They include:

- a further period of surface or near-surface storage, probably for some decades and maybe longer, to allow heat production to dissipate, followed by deep disposal of spent fuel;
- a further period of surface or near-surface storage, followed by reprocessing and disposal of the HLW, perhaps after vitrification;
- early reprocessing of spent fuel, followed by vitrification of the HALW and storage of the resulting VHLW;
- early reprocessing of spent fuel, followed by deep disposal of VHLW;

- construction of deep storage facilities (available for either spent fuel or VHLW), to be converted to disposal if and when there is sufficient confidence to do so;
- partition and transmutation of spent fuel or highly active waste, followed by interim storage or final disposal.

All would involve a number of steps, and all would involve significant public participation in decisions at each step. Building and maintaining public confidence throughout an inevitably long process would be a considerable challenge, undoubtedly with profound implications for whatever approach or combination of approaches was followed in different countries.

Deep disposal of radioactive materials

Whether or not spent fuel is reprocessed, a decision will be required in due course as to whether to dispose of the more radioactive waste – the spent fuel itself or the separated HLW – in a deep disposal facility or to continue to store it on the surface or in a monitorable underground facility from which it can be retrieved. Similar decisions are applicable to long-lived intermediate-level waste. In fact, this decision is not symmetrical: disposal of waste is by intention irreversible (although in practice retrieving it would doubtless be possible if the requirement were severe enough), but continued storage does not preclude future disposal.

Building public and political support for siting a deep repository would represent a major challenge in most countries, an issue that has been considered in Chapter 2, on public perceptions and decision-making. A deep disposal facility must also address the following requirements, as well as matters of cost:

- prevention of leakage of unacceptable levels of material to the surface over long periods of time;
- prevention of criticality within the facility itself;
- prevention of appropriation of contents for weapons proliferation purposes.

The second requirement, the prevention of conditions in which nuclear fission could occur, is a matter of preventing the accumulation of fissile materials such as uranium-235, plutonium-239, plutonium-241 or neptunium-237 to a dangerous level of concentration and in the presence of water. It is generally thought that this is not a challenging technical issue in view of the nature of the wastes involved, although detailed repository design will have to demonstrate this. The issue of preventing proliferation will be considered in more detail later. This section will focus on site integrity and the prevention of leakage into the natural environment.

Underlying containment of radioactive materials in deep repositories is the 'race' between two forces of nature. The radioactive material is decaying over time, while natural forces tend to disperse the material through the environment. The most likely way that significant amounts of material could return to the surface is through the movement of water, although there are also concerns about gaseous leakage, owing to the build-up of gaseous radionuclides such as radon or krypton or of methane and other gases from composting organic material in the repository. Some disposal concepts, for example in Sweden and Finland, involve an intention to contain the material by putting the waste in containers made of a material such as copper and placing these in buffer material with low groundwater flow. Others allow for leakage from the facility and rely on the geosphere to isolate the material from the environment.

Fundamental to the deep-disposal philosophy is the 'multi-barrier approach'. Several different barriers are placed between the waste and the surface, with the aim that the radioactive components of the waste will not spread from the site until their activity has decayed to, or close to, background levels.

The barriers are usually classified into three kinds: physical, chemical and geological. Physical barriers include the packaging of the material. Most spent fuel is in the form of a ceramic, and liquid HLW is converted into glass blocks and packaged in stainless steel. Intermediate-level waste, such as the cladding materials in which the fuel has been held in the reactor core, is usually placed in steel barrels and backfilled with a cement grout. These physical barriers will slow the rate at which water, when it has invaded the site, will gain contact with the radioactive material itself. However, attention must be paid to the possibility of gas breaching the barriers. The construction of the repository itself, and backfilling the whole site with concrete when it is full, also acts as a physical barrier.

Chemical barriers consist of, for example, the use of cement. As water dissolves the cement, the water becomes strongly alkaline. Most of the radioactive species in the waste are rather less soluble in alkaline conditions than in neutral or acidic ones. The choice of bedrock can also be important. Some rocks are especially effective in absorbing the nuclear material as it passes through – in effect, the radionuclides 'stick' to this rock rather than move with the water in which they are dissolved as it flows away from the repository.

The main geological barrier is the several hundred metres of rock between the repository and the outside world on the surface. Sites will also be chosen where there is very little water and where what water there is moves only very slowly. Political considerations, however, may limit the range of sites that are available in practice.

Arguments for and against early disposal

For some time there has been a strong international technical consensus within the nuclear industry and its regulators that underground disposal is the more attractive approach, at least for highly radioactive and long-lived wastes. But it has failed to command political and public support in many countries, and many commentators argue that important technical problems remain to be overcome.

Besides stability, which, it is claimed, is conferred by the multiple barriers discussed above, the reasons adduced for the early disposal of waste include:

* an ability to deal with the problem in this generation rather than bequeathing it to the future;
* less risk of deliberate or accidental damage than is the case with surface storage;
* lower worker doses;
* a smaller number of sites.

Of these, the first two are generally regarded as the most important.

The second issue has become especially prominent since the terrorist bombings in the United States in September 2001. If it should appear more likely than was previously assumed that deliberate attempts may be made to damage waste stores, then probably underground storage, or underground disposal, will become correspondingly more attractive. However, it can be argued that relatively shallow underground storage, say at a depth of 20 to 40 metres, might be sufficient for this purpose without having to go to the 500 to 1,000 metres envisaged for deep disposal. The importance of conditioning the waste to ensure passive safety when in storage has also been emphasized by the events of September 2001. It should be noted that as long as nuclear power continues to be used, there will always be a number of surface sites holding some spent fuel, immediately after discharge from the power stations.

Discussion of the first point is often couched in terms of 'intergenerational equity': as the waste arises from the activities of people alive today, who receive most of the benefits of using nuclear technology, it should fall to this generation to dispose of it. This argument is questionable, on two distinct grounds. The first is that people alive in the past four or five decades are also the people who have provided the investment necessary to develop nuclear technology from a theoretical possibility to a fully developed potential energy source. The benefit of a mature technical option, including the benefit of avoided fossil fuel use and greenhouse gas emissions in the past and future, will accrue to future generations as an offset against the radioactive waste legacy. Balancing these considerations against each other is not straightforward. Second, the argument for immediate action on intergenerational grounds holds force only if it is assumed that a final solution is available and can be demonstrated to the satisfaction of relevant stakeholders. Committing people of the future to a 'solution' that did not work would be the antithesis of sustainable development.

Opponents of deep disposal, supported by the views of several geologists, engineers and chemists, argue that at present (and maybe in principle) the understanding of geology is not sufficient to offer guarantees that radioactive material will not escape from the repository more rapidly than calculated and reach the surface before its radioactivity has decayed to acceptable levels (assuming that there are 'acceptable levels'). They question whether the action of constructing the repository might change the fundamental nature of the geology, for example during excavation, and lead to potentially unpredictable water flows through the cracks created. They point out that the access shaft represents the material's potential escape route to the surface. They also question whether the understanding of the chemical behaviour of the large number of radioactive species in highly active wastes is sufficient to predict how they will behave as they migrate from the site. Once material has escaped from the site, they argue, it will be extremely expensive to retrieve the bulk of the radioactivity, and impossible to retrieve it all. Continued storage of the material in a monitorable and retrievable condition would therefore seem to be the only possible way forward. This argument may be combined with a call for the production of radioactive waste from power and military sources, though generally not from medicine and other non-power uses, to cease as soon as practicable.

Whatever the relative merits of these arguments, there would seem to be a clear moral obligation to make a start towards a long-term underground disposal strategy that, as mentioned earlier, might have three components:

*   the identification, storage and packaging of all wastes;
*   the development of an infrastructure for implementing a long-term plan;
*   the accumulation and preservation of funds to pay for it.

However, the support of communities for this strategy cannot be taken for granted, even in those cases where the power station itself has local support. Communities may, for example, seek assurances that the waste would be moved to a final disposal site within a certain time as a price for acceptance.

Natural analogues

There are a number of examples from nature that suggest that the containment of radioactive materials for very long times is possible, at least in principle. Perhaps the most famous is the 'Oklo phenomenon'.[17] In the early 1970s, French scientists noticed that some uranium samples from the Oklo mine in Gabon in West Africa had an abnormally low amount of the isotope U-235, the active isotope in thermal fission reactors. Because U-235 has a shorter half-life than U-238, the major component of natural uranium, its concentration has been falling over the 4.5 billion years since the earth was formed. Some 1.7 billion years ago, the proportion of U-235 in natural uranium was about three per cent, similar to the enriched uranium used in most modern reactors. Water filtering down through crevices in the rock could act as a 'moderator', as required in most modern nuclear reactors, slowing down neutrons and thus allowing a nuclear reaction to take place.

It is believed therefore that natural reactors could have functioned intermittently for a million years or more, until the U-235 became too depleted. Sixteen such 'fossil' reactors have been identified in the Oklo region. Once the natural reactors burned themselves out, the highly radioactive waste they generated was held in place deep under Oklo by the granite, sandstone and clays in the surrounding area. In the subsequent 1.7 billion years, the plutonium (some four tonnes of which was produced by the reactors) and minor actinides have moved less than three metres from where they were formed, although they have decayed away to stable elements. It is less easy to determine the fate of some of the more mobile or volatile fission products.

Although the 'Oklo phenomenon' may be the most widely known of the natural analogues, there are others that geologists believe may be more relevant in their wider implications. The Cigar Lake uranium deposit in Saskatchewan, Canada contains the world's richest known uranium deposits, with an average 14 per cent of uranium but reaching 55 per cent in some areas. Its geology is relatively simple. It lies at a depth of over 400 metres and occupies a lens-shaped area 2,000 metres long, 100 metres wide but only one to 20 metres

---

[17] Knight (1998).

deep. It is almost totally enclosed in a clay-rich envelope. The ore was laid down some 1.3 billion years ago, but there is no geochemical or radiometric evidence of its existence on the surface, despite the fact that most of the intervening rock is fractured and bears considerable amounts of water.[18]

Cigar Lake is regarded as the most complete natural analogue for a repository, having been successful in containing uranium for over one billion years in conditions with many similarities to repository concepts. Several other natural analogues in regions in which uranium is mined have also been investigated and characterized, notably the 1.8 billion-year-old deposits at Alligator River at Koongarra in Northern Territory, Australia.

Although useful in many ways, the study of natural analogues has its limitations. For example, these natural phenomena tend not to include the range of different chemicals associated with spent fuel or high-level waste, nor do they have entry shafts that have been excavated in the natural rock formation and may be a means of escape for the radioactive materials.

The philosophy of the disposal problem

It is generally agreed that an 'acceptable' waste management strategy would not only have to satisfy technical requirements (that is, a suitable safety case could be made) but also be acceptable in public and political terms. But building public confidence may be difficult. This topic has been considered in depth in Chapter 2, on public perceptions and decision-making.

There are two possible reasons for arguing that a particular waste management proposal, say for a deep repository, is not acceptable on technical grounds. First, the degree of understanding of the site in question, or the techniques to be used, is not yet sufficient. More research might be needed to characterize the water flow through the area, for example, or to understand the interaction of some of the radioactive species with various kinds of rock. Furthermore, there is a need in some cases to establish agreement on what a particular observation might mean. Sometimes more data can result in less certainty, for example if those data contradict the predictions of the 'best fit' model that has been developed from previous investigations. A degree of uncertainty will inevitably remain.

However, in refusing permission to construct a facility for this reason, the authorities would not be precluding the possibility that a proposal could be made in the future, when the science was sufficiently advanced and the re-

---

[18] Smellie and Karlsson (1996).

maining uncertainties were deemed to be manageable. All the same, advocates of deep disposal might argue, natural analogues show that, in principle, radioactive material can be retained in geological formations for periods much longer than is necessary for complete nuclear decay. Moreover, this occurs even without elaborate safety precautions, although there is clearly a gap between the possibility of retention in principle and the ability to guarantee containment at a particular site.

The second possible argument for opposing an application to construct a repository is quite different. Its premise is that because the timescales involved are so long, it is not possible in principle to guarantee that measurable amounts of radioactivity will not leach to the surface at some time. The argument here is that no amount of research will overcome the basic problem of induction – that we cannot be certain that the future will be like the past and that the distant future is less predictable than the immediate future. If this is true, and if radiation is regarded as unacceptably dangerous at all dose rates, then the only course of action is to stop producing waste and to monitor actively the waste that has already been produced in order to ensure it is contained as effectively as possible. Presumably, this would apply to waste from medical and other non-power uses of radioactive materials as well as to that from nuclear energy and military uses.

It is important to distinguish between these two arguments. To an extent, the anti-nuclear movement is raising fears along the latter line of reasoning, while the nuclear industry and its supporters try to find answers to the former (and indeed many would argue that answers are already available, although this is of course hotly disputed by other scientists). In fact, even if the second line of argument were accepted, it does not follow that deep disposal should not be pursued, as it might still appear to be the most likely approach for fulfilling the requirements of a waste management strategy. This issue is of major significance when it comes to considering future arisings of nuclear waste.

If it is agreed that no further nuclear power development should take place until a final waste management policy is demonstrated to be effective and that no such demonstration is possible in principle owing to the timescales involved, then it follows that no further nuclear development should ever take place. (This would apply unless a more radical approach such as partition and transmutation become feasible, which would lead to a significant reduction in the lifetime of most of the waste.) The argument is not weakened significantly by the observation that future waste arisings will be much smaller in volume than historical wastes. For those whose opposition to nuclear power is fundamental in nature, this represents a very powerful and useful argument.

It can be countered that the same considerations are applicable to fossil fuel-fired power stations and their pollutants, perhaps especially carbon dioxide. The consequences of an overnight closure of all fossil fuel plants, or even of an immediate policy to prevent the building of these plants, would be far-reaching, especially in developing states such as China and India. Those putting forward this argument hold that cost-benefit analyses must be applied across the energy field, including the discussion about future nuclear development, in preference to applying absolute prohibitions to any individual process or material.

The immediate future

Despite a series of proposals going back some years, many countries have made relatively little progress in finding a publicly and technically acceptable route to manage at least the more radioactive of nuclear wastes. Nor does it seem likely that they will find a solution in the next decade or so. But even if the basic science is sufficiently well understood, and some scientists and engineers vigorously dispute this, it would probably take a long time to identify and characterize a suitable site in order to make a safety case. It would then presumably be necessary to gain the trust and confidence of the local communities involved, a matter that has been notably difficult in many, although not all, countries where attempts have been made or are being made to site waste repositories.

Some commentators also question whether appropriate political structures exist in all countries for taking these decisions. It can be argued that in the absence of other pressures such as legal obligations, it may never be in the interest of the government of the day to set in train the construction of a final waste repository. This decision would be both controversial and expensive, and there would always be the apparent alternative of surface or near-surface storage, mainly at site of origin of the waste, perhaps coupled with continued 'research' into the basic and applied science of waste disposal facilities.

It would seem to follow, then, that a period of interim, monitorable and retrievable storage of the wastes in question is inevitable in these countries, as a 'solution', or even a workable plan, is unlikely to be available soon. The observation that in some countries a shortage of space for spent fuel could result in the closure of operational reactors unless new stores are built in the near future adds to this conclusion.

One can speculate that some of the key uncertainties now facing energy strategists may become at least somewhat clearer in the next few decades. There might be a better understanding of climate change and a better appraisal of the practical potentials of renewable sources of energy and of partition and transmutation of radioactive materials, for example. A partial resolution of these issues would presumably lead to a clearer view of the need for nuclear power, and thus of the scale of future waste arisings. This in turn would have a profound influence on waste management strategies, because very different approaches to waste management and related issues such as reprocessing might be appropriate in a world in which the nuclear industry was declining or expanding rapidly. However, it is unrealistic to imagine that the future, especially where it refers to human society, will ever be 'knowable'. A scenario approach, with as much flexibility as possible for responding to many different futures, will remain the most attractive option.

The issue may arise, then, as to whether replacement interim storage (either on the surface or underground) should be designed and constructed with a longer potential operational lifetime – say 100 to 200 years – than is often discussed at present. This would allow for a more flexible response to future scientific, technical and economic developments. Crucially, in the eyes of many commentators, this approach would not foreclose future options in the way that an early move towards disposal would.

However, even if it should prove to be technically possible to pursue this option, care would have to be taken that it would not become, or be perceived to become, an excuse for inaction. It is likely that affected communities would accept the strategy only if it were accompanied by clear evidence that progress was being made towards a longer-term solution or approach and that retention of the material on site for long periods would be undertaken only as a 'last resort'. It is of course a major question as to what kinds of promise could be offered or accepted to this end.

Deep storage

In recent years the concept of deep storage, incorporating the possibility of reversing it should circumstances dictate, has gained ground, a trend that may accelerate after the terrorist attacks of September 2001. A monitorable and retrievable store could be built underground. Waste would be put in it, and the condition of the waste would be monitored through the lifetime of the store and, if necessary, for a period afterwards. If and when there was confidence that the site could be sealed, and therefore converted into a disposal facility, this could be done. If, however, the monitoring were to demonstrate a leakage of materials before it was expected, or if better approaches were developed, the waste could be retrieved.

This option looks expensive, however. Some commentators have suggested that it would be a 'Trojan horse' and that, once built, the momentum towards sealing it, independent of any scientific assessment, would be irresistible. The site requirements for deep stores and deep disposal facilities may be quite different. Thus a deep store would need to be resistant to hazards such as roof falls and flooding. This would be less important for disposal facilities, which are likely to be backfilled with suitable materials. Storage facilities would not require a very long-term safety case or perhaps need to be set at the same depth as a final repository. Furthermore, creating access for monitoring might also create a potential pathway through which radionuclides could migrate from the site, and a prolonged aerobic environment could be problematic for the integrity of packaging.

Economic aspects

As noted earlier, uranium is a very concentrated energy source, and the volumes of waste being produced are, by industrial standards, modest. Waste management costs thus represent a small fraction of the total costs of nuclear power generation, which are dominated by the initial construction costs of the reactors. Furthermore, many of the costs of waste management (especially those associated with decommissioning nuclear facilities) are incurred quite late in the economic life of the facility – indeed, a significant proportion will not be incurred until after closure. Even with discounting at a modest rate of two or three per cent, those costs will represent a proportionately lower present value. (There is a lively debate about the appropriate discount rate to use; some commentators argue that a rate closer to commercial rates of return might be appropriate, while others argue that to value properly the future environment, no discounting should be applied at all. This may seem to be a somewhat arcane economic argument, but it has enormous implications for the quoted size of liabilities in the nuclear field, and indeed in other heavy industries as well. A sum of $1 billion, discounted at two per cent over 100 years, would have a present value of $138 million. Discounted at eight per cent, the figure would be $4.5 million. Without discounting, it would of course remain at $1 billion.)

There will come a point at which the costs become 'real' – the liabilities must be discharged – and in absolute terms they could be very large indeed. Large amounts of radioactive waste have already been produced. More will arise when existing facilities are decommissioned, whenever this should happen. There are serious questions in many countries about whether sufficient resources have been put aside to deal with present and committed arisings of nuclear waste from existing facilities or whether the taxpayer of the future will be landed with significant liabilities. These questions are given immediacy by the burgeoning decommissioning and waste liabilities associated with older reactor systems in many countries. Provisions for decommissioning these facilities were built up during their operating lifetimes, but the sums have in many cases proved to be inadequate. In some cases, notably the United Kingdom, the provisions themselves were reinvested in new electricity generation or distribution projects or in nuclear fuel cycle facilities. Some of these projects were subsequently sold to the private sector. However, the part of the proceeds from the sale that represented the waste and decommissioning provisions was often not preserved in a dedicated fund, while other projects, which had been financed from the industry's cash flow, failed to make expected returns. As a result, the shortfall had to be made up within the short remaining lifetimes of the facilities, or the financing of decommissioning and waste management was left to future taxpayers.

By contrast, in the United States the industry has been paying a levy to the federal government since 1983 towards funding a disposal programme for spent fuel, but progress in implementing the policy, which includes proposals for a major facility at Yucca Mountain in Nevada, has been slow. In these circumstances the nuclear industry has been denied the option of reinvesting the sums involved, without benefiting from a solution to the short-term or long-term problems associated with waste management. In any case, there could be no guarantees that sufficient provision had been made, and, as discussed earlier, some estimates suggest that there will be a shortfall even in the US case.[19]

Ensuring that sufficient provision will be available to discharge liabilities when they become due without unduly penalizing the industry is therefore a complex and urgent issue, especially where nuclear utilities are in private ownership. Although the costs of waste management will be a small proportion of nuclear costs when averaged out over the entire output of the power stations, the absolute costs to future generations if insufficient provision is made could be enormous. Experience thus far suggests that what might appear to be a 'prudent' estimate of management costs can eventually prove to be rather optimistic. A resolution of this matter, in such a way that there can be confidence that funds will be available at the appropriate time whichever waste management approach or approaches are eventually followed, is likely to be an important prerequisite of a major expansion of nuclear power.

---

[19] Loux (1998).

This issue is not unique to the nuclear industry. The decommissioning liabilities of other energy facilities, such as coal mines or oil rigs, and indeed of plants in other heavy industries such as chemicals, can also be very significant. (An example is the problem of subsidence and contamination of water caused by closed coal mines.) But they are not in general dealt with by creating segregated decommissioning funds. However, some commentators argue that the issues are more serious in the case of nuclear power, given the timescales, the political issues and the question marks over the long-term future of the industry itself.

An international approach

In recent years, attention has been paid to the possibility of developing a small number of regional repositories that could take radioactive materials from a number of countries. Thus Russia is offering to import spent fuel for long-term storage, and there has been interest about possibly using isolated areas in Australia for waste disposal.

From time to time there have been suggestions that China might develop a site in the Gobi Desert for this purpose, although the Chinese have no plans to open up a repository that might be built there for international use. In July 2001, Kazakhstan announced that it was considering importing and storing low-level nuclear waste on its territory. The reasons for this attention are numerous. From the point of view of the potential host country, these proposals may represent a considerable source of income, to be measured in tens of billions of dollars. For customer countries, international repositories could offer a more cost-effective or, cynics would say, a more politically acceptable solution to waste management problems at home.

It should be noted that there are many activities, some unrelated to power production or the military, that give rise to the production of radioactive wastes. A number of countries have only one or two operating nuclear power reactors, at least at present: Argentina, Armenia, Brazil, Kazakhstan, Lithuania, Mexico, the Netherlands, Pakistan, Romania, Slovenia and South Africa. Over 30 countries with no nuclear power reactors have research reactors or critical assemblies (either under construction, operating or closed), among them Bangladesh, Belarus, the Congo, Georgia, Ghana, Jamaica, Latvia, Nigeria, Syria and Vietnam.[20] It can be assumed that every country in the world uses radioactive materials for medical and other non-power, non-military uses.

---

[20] IAEA (2001).

The potential dangers of these sources of radiation should not be underestimated. For example, in 1987 a canister of caesium-137 that had been left in an abandoned radiotherapy machine in Goiania, Brazil was found by scavengers and prised open. As a result, 244 people received significant radiation doses, four of whom died within a week. These materials and their resulting wastes will clearly need safe management for considerable periods of time. The issue of radioactive waste management has been of considerable importance in, for example, negotiations over accession to the European Union for some central European countries.

Some commentators argue that it seems to make little sense, on safety, environmental or economic grounds, for every country to develop facilities for managing all categories of nuclear waste, no matter what the state of their nuclear expertise and experience. They hold that a deep geological facility would be likely to cost well over $1 billion, no matter how small the volume of wastes to be disposed of, and it is inconceivable that each country with wastes could afford this. For these countries, it is argued, shared repositories appear to be essential, especially if deep disposal should become the standard long-term management approach. For other countries, complex geology or intense land use might justify pursuing international options. Others argue that international repositories for managing bomb-grade nuclear materials liberated by the destruction of nuclear weapons could enhance global security and reduce proliferation risks, by simplifying the task of preventing the malicious use of these fissile materials.

Advocates claim that the main scientific advantage of a global or regional choice of geological environments would be not absolute levels of safety but rather the confidence with which future safety could be predicted. An international disposal facility that had demonstrably high levels of safety and strict safeguards and was open to the scrutiny of all participating countries (perhaps under the auspices of the IAEA) could have significant environmental and security advantages. There seems to be no reason why national and international programmes could not coexist.

However, opponents of regional depositories claim that they present a number of difficulties. First, the option to offload domestic responsibility for managing waste would weaken the pressure on producing countries to reduce the amount of waste being created. This view is often coupled with the observation that the moral responsibility for managing wastes should lie with the country that produces the waste. Far from seeking to export waste to developing countries, it is argued, developed countries should be prepared to accept waste from those countries less able to deal with it. Second, concerns have been raised about the technical capability of some of the volunteer host states to manage toxic material safely for long periods of time. The potential hosts sometimes counter this argument by saying that the resources generated by the scheme could be used in part to enhance their management of their own wastes, thus improving overall safety. However, a very high level of international scrutiny of the proposed facilities, at all stages from design to closure, would seem to be necessary, possibly to the extent of infringing the sovereignty of the country involved. Third, the development of international repositories would involve the transportation of large amounts of radioactive material for considerable distances, depending on the siting of the facilities in question.

Whatever the attractions or otherwise in environmental or economic terms, it is clear that as international law stands now, it would be very difficult to establish a network of a small number of international radioactive waste disposal sites. Export to African, Caribbean, Pacific or Antarctic countries is expressly banned, while the creation of regional centres in OECD countries would be likely to meet major political difficulties. International law and political opinion have for some time been moving towards a position according to which the exportation of hazardous waste should take place only in those cases where the country of origin is technically or financially incapable of disposing of that waste safely – that is from developing countries to the developed world, not the other way round or between developed countries.

Partition and transmutation

A more radical, but as yet undemonstrated, approach to managing highly radioactive materials has become known as 'partition and transmutation' (P&T). This is designed to convert radioactive atoms either into stable ones or into ones with shorter half-lives. One version, which has been the subject of research at Los Alamos in the United States, is known as accelerator-driven transmutation of waste (ATW).[21]

P&T has three steps. The first involves separating the spent fuel into three fractions:

* unused uranium, for reuse or storage;
* short-lived fission products, such as caesium-137 and strontium-90 (with half-lives of about 30 years) that dominate the heat production in spent fuel in the short term, for separate storage until they decay to stable forms;

---

[21] Venneri et al. (1998).

• actinides such as plutonium and americium, along with long-lived fission products (such as iodine-129, whose half-life is 16 million years, and technetium-99, whose half-life is 210,000 years) which are responsible for the radioactivity of the spent fuel in the long term.

It is envisaged that this step will be carried out using a non-aqueous (pyrochemical) process. Sufficient uranium (99 per cent) must be separated from the spent fuel in order to ensure that in the subsequent process no new actinides are produced. However, the techniques for separation look to be some way from demonstration on commercial scales.

The second step is to bombard the third, long-lived fraction with a beam of neutrons, produced in the ATW process by a high-power proton linear accelerator. This fraction is 'burned' in a liquid lead–bismuth-cooled facility. This step produces heat that can be converted to electricity, some of which (up to 15 per cent) would be required to power the accelerator, the rest of which could be sold commercially. Another option might be to burn up this material in a fast reactor. For technical reasons, thermal reactors appear to be far less appropriate for these purposes.

The final step would be the storage and/or disposal of the more radioactive products.

The advantages of P&T, according to its advocates, are several. The bombardment of the residue, it is claimed, would result in the destruction of over 99.9 per cent of the actinides and over 99.9 per cent of the technetium and iodine. The time horizon for waste repositories would be reduced from hundreds of thousands of years to a few hundred years, as the long-term radiotoxicity of the waste would fall by a factor of 1,000. (However, other commentators question whether the process will be this efficient, and suggest that certain long-lived fission products, notably chlorine-36, tin-126 and caesium-135, could not be transmuted in the way suggested. In addition, the ease with which the products of transmutation could escape from a repository, which varies from substance to substance, has to be taken into account.)

Heat production within the residual longer-lived waste (the third fraction described earlier) would be much lower. This is because the main heat-producing isotopes would have been separated, making repository design easier, although of course the heat-producing isotopes would require storage until they have decayed. The volume of material for disposal would be much smaller than the volume of the original spent fuel, as the major component, uranium, would have been removed. At no point during the process would plutonium be separated from highly active waste products. Accelerator-based P&T might be particularly appropriate for treating mixed or poorly categorized spent fuel, which might be difficult to burn up in a reactor because of unknown criticality characteristics.

It is further suggested that P&T could become the basis of an alternative approach to power production, which uses sub-critical assemblies. This equipment does not involve self-sustaining ('critical') fission reactions, but instead depends on an external source of neutrons that could easily be switched off in the event of problems. Control rods would be much less important than in critical technologies such as power reactors, and would be functionally separate from the source of neutrons. (In a power reactor, control rods affect the very neutrons that keep the reaction going.) Nonetheless, the decay heat in the transmuted materials would be considerable, so loss-of-coolant accidents would have to be prevented.

A number of other points should also be considered. Although the basic science may be possible, a major R&D programme, estimated at several billion dollars, would be required before statements could be made about the technical or economic feasibility of the P&T technique. It would not be appropriate for relatively dilute wastes (ILW) because the process of neutron bombardment would create radioactive materials from the bulk of stable material in that waste. Although some longer-lived radioisotopes could be transmuted into shorter-lived ones, the opposite could happen too, and the effect of protons and neutron beams on packaging and containment materials must be considered.

Furthermore, P&T would not be appropriate for existing wastes, especially not vitrified high level waste, unless some method could be developed for breaking down the vitrification and releasing the individual radioisotopes for the partitioning process. The potential benefits of slowing down the vitrification of highly active liquid wastes, in the hope that this waste could eventually be partitioned and transmuted, would have to be set against the extra risks of environmental contamination incurred by retaining the highly active waste in liquid form. This proposal has few, if any, advocates.

P&T would seem to make sense only in the context of a continuing world nuclear power industry. Nevertheless, advocates of the concept argue that should global use of nuclear power increase significantly over the next decades, P&T could be an attractive way of managing the most highly radioactive materials acceptably, perhaps within two decades if sufficient effort were expended. Other commentators are rather more sceptical.

Nuclear weapons proliferation

Nuclear weapons predate civil nuclear power. It took less than three years, using 1940s technology, from the first demonstration that nuclear fission was a practical possibility to delivery of the first atomic weapons. It is therefore highly improbable that there are any purely technical steps that would prevent a determined state from developing at least fission weapons based on uranium (which can be developed without the use of a nuclear reactor). Modern developments such as laser-induced enrichment may make the production of nuclear weapons easier.

It would appear unlikely that a state such as South Africa, which has renounced nuclear weapons and destroyed its capabilities, would restart a military programme simply because it had decided to build more nuclear power stations. However, there are at least two ways in which the promotion of a civil nuclear industry could aid a state in developing a weapons capability or offer opportunities to sub-state organizations to do so. First, materials and equipment used in nuclear power production could, in principle, be diverted to military use. Second, a nuclear programme, including perhaps reactors used for research, for producing isotopes for medical or industrial use or for use in non-destructive testing of materials rather than (or as well as) power reactors, would create a body of personnel and expertise with capabilities in nuclear science and engineering.

It can be argued that a civil nuclear industry based in the developed world can play, and has played, a role in slowing the spread of nuclear weapons. The Nuclear Non-Proliferation Treaty (NNPT), for example, offers help in developing civil nuclear power to states that undertake not to divert the resulting materials or skills towards a military programme and that accept verification procedures from the international community. In effect, aid in developing civil nuclear energy is being used as a 'carrot' to encourage countries not to develop weapons. These procedures might include inspection visits and other monitoring activities. Failure to abide by the provisions of the treaty could result in an international enforcement response, including export controls, sanctions, embargoes and legal action.

The NNPT has largely been successful. The five nuclear weapons states, the United States, Russia, the United Kingdom, France and China, had all tested nuclear devices by 1964. However, little if any progress has been made, at least until recently, towards disarmament among these states as required by the treaty. Since 1964, only two other countries, India in 1974 and Pakistan in 1998, have exploded nuclear devices. Israel is believed to have a nuclear capability. These three countries are not signatories to the NNPT. Iraq and North

Korea are believed to have embarked on weapons programmes despite being signatories of the treaty, although the regime of inspection associated with membership did give some warning.

Another important agreement is the Convention on the Physical Protection of Nuclear Material, which entered into force in 1987 and has been ratified by over 30 countries. The convention provides for certain levels of physical protection during the international transportation of nuclear material. It also establishes a general framework for cooperation among states in the protection, recovery and return of stolen nuclear material. In addition, it lists serious offences involving nuclear material which parties are to make punishable and for which offenders will be subject to a system of extradition or submission for prosecution.

It should be noted that the threat of rogue states or terrorists developing or obtaining nuclear weapons is not the only risk associated with radioactive materials. It might also be possible, for example, to contaminate a conventional weapon with radioactive material and create a 'dirty' bomb or to release spent fuel into drinking water supplies, although the use of chemicals or biological agents would probably be simpler and more effective.

Separated plutonium

As noted in Table 4.2, by 1995 the world held an estimated 1,240 tonnes of plutonium, of which some 250 tonnes were classified as 'military'. There were 190 tonnes of separated 'civil' plutonium (in storage or in fuel cycles) and about 990 tonnes of civil plutonium in unreprocessed spent fuel.[22] In all scenarios that appear credible, the use of plutonium for fuel in the next few years will account for only a small fraction of this material, and for a smaller amount than is at present being separated out during reprocessing.

Weapons-grade plutonium contains a high proportion (at least 93 per cent) of the isotope plutonium-239, and has usually been produced by keeping the fuel (containing uranium-238) in the core or dedicated reactors for a relatively short time, typically about three months. Generally, a dedicated reactor is used for the purpose. Plutonium-239 is a particularly effective nuclear explosive.

The plutonium that exists in normal reactor spent fuel, which will typically have been in the reactor core for about five years, contains significant proportions (up to 50 per cent) of higher plutonium isotopes (mainly 240, 241 and

---

[22] IAEA (1997).

242). The presence of these higher isotopes may make it more difficult to produce a nuclear weapon – there is some dispute even on this point – but by no means impossible. In the words of the US Department of Energy, 'Virtually any combination of plutonium isotopes can be used to make a nuclear weapon. Not all combinations, however, are equally convenient or efficient.'[23] There remains some debate about how great are the differences in efficiency between military- and civil-grade plutonium in producing weapons, but some calculations suggest that up to 40 per cent more civil plutonium is required than military plutonium to make a device of the same yield.[24] A bomb made from civil plutonium may also have a greater tendency to 'fizzle', although the yield can still be large (equivalent to one or two thousand tonnes of TNT).

One tonne of military plutonium is sufficient to create 200 to 300 warheads. The dismantling of large numbers of warheads as a result of the disarmament process has released considerable stocks of military plutonium, which have to be managed.

Managing plutonium is, and will remain, an important task. This will be true whether the use of nuclear power should decline, continue at about present levels or expand considerably.

Options for plutonium management

There are two broad options for plutonium management:

- immobilization (fixing the plutonium in a physical matrix that makes it impossible, or very difficult, to separate the plutonium out and/or mixing or surrounding it with highly radioactive materials so as to make it unattractive for proliferation purposes);
- utilization and consequent destruction (in fast reactors, in high-temperature gas reactors (HTGR) or in mixed oxide fuel or through transmutation).

In the 'once-through' cycle, the intense radioactivity of the fission products in the spent fuel in effect renders the plutonium too dangerous to any organization or state without reprocessing and remote-handling facilities. However, as mentioned earlier, this radioactivity dies away significantly over a few centuries, after which the spent fuel may well be an attractive source of plutonium for potential weapons production, and perhaps of other proliferation materials

---

[23] US Department of Energy (1997).
[24] Makhijani (2001).

such as neptunium-237. This creates tension with one of the basic assumptions in repository design, that records will be lost, and therefore that future generations should not be required to maintain vigilance against accidental damage or deliberate misappropriation of materials.

Separated plutonium could also be immobilized, by fixing it in a glass or ceramic matrix. Barker and Sadnicki argue that creating ceramic matrices may be the most attractive way forward.[25] Two ceramic options are a MOx waste form or a purpose-designed ceramic.

The authors considered three approaches to external radiation:

- no radiation barrier (the proliferation deterrent being the bulk and size of the container and the need for dissolution and separation from the ceramic);
- a spent fuel radiation barrier;
- a vitrified HLW radiation barrier.

The particular option to be pursued might vary, depending on the particular characteristics of the materials in question, which might differ from country to country.

Immobilization in this way would make sense only against a background assumption that plutonium was to be regarded as a waste, and therefore that the dangers of storing it in its present form outweigh any likely benefit of retaining it as a potential reactor fuel that might be both commercial and acceptable in the foreseeable future. Barker and Sadnicki argue that in the British context, the use of MOx in reactors could not be justified on a commercial basis. (In the longer term, it would be possible to move towards a 'plutonium economy', if this were thought necessary, with a relatively small amount of plutonium and a programme of breeder reactors. But this would take several decades to implement.) Their argument seems to assume no major increase in nuclear power output in the next few decades. It is not clear that an approach whereby a proportion of plutonium in a store were immobilized while the rest was retained for possible future use would make sense, as the proliferation implications of one tonne of plutonium held at a single site or 50 tonnes would seem to be similar.

Utilizing plutonium succeeds in reducing the total amount of plutonium in existence in comparison to a non-utilization approach (assuming that the total amount of nuclear-generated electricity is the same). However, with the exception of the partition and transmutation approach discussed earlier, it requires the

---

[25] Barker and Sadnicki (2001).

separation of plutonium. In France, for example, there are hundreds of shipments of plutonium oxide and fresh MOx pins, assemblies and fabrication residues each year. Although no weapon has ever been made by diverting this material, the technology to produce a crude nuclear device is not enormously complex once the perpetrators have access to fissile material. Furthermore, as mentioned earlier, one can imagine that a conventional bomb laced with plutonium (or any other toxic material) might be an attractive weapon for potential urban terrorist use in view of the levels of public concern about radiation in some countries.

Mixed oxide fuel typically contains about 3 to 10 per cent plutonium, in oxide form, mixed with uranium oxide in a ceramic base. MOx has been used in reactors since the 1960s, and countries such as France, Germany and Switzerland have been using it in commercial nuclear power plants for many years. There are 35 commercial nuclear power reactors currently loaded with MOx fuel in Europe: 20 in France, 10 in Germany, three in Switzerland and two in Belgium. Up to 70 reactors worldwide, in countries such as Japan, Russia and the United States, are reportedly scheduled to use the fuel by 2010. Typically, a reactor loaded with MOx fuel will use approximately 30 per cent MOx and 70 per cent conventional reactor fuel.

At present, the rate at which the world is using MOx is slower than the rate of separation of plutonium through commercial reprocessing, so stockpiles of separated and stored plutonium are growing. It should be noted too that although a proportion of the pre-existing plutonium is destroyed during the use of MOx fuel, some of the uranium oxide is transmuted into fresh plutonium. So far, the reprocessing of MOx spent fuel, which still contains significant amounts of plutonium, has not been demonstrated on a commercial scale.

However, it is feasible to increase the rate at which plutonium is being used in thermal reactors, in at least four ways. The first would be to convert more existing reactors to run on MOx. The second would be to increase the proportion of MOx fuel in the fuel mix being used by existing reactors. The third would be to build reactors dedicated to burning MOx and capable of being loaded 100 per cent with the fuel. (It should be noted once again, however, that the use of MOx does not necessarily reduce the total amount of plutonium in existence, as new plutonium is produced from the uranium oxide present.) The fourth would be to develop fuel forms in which the plutonium oxide is held in an inert matrix, with the result that no fresh plutonium is produced from the uranium oxide. All of these options would require careful safety cases and considerable development work.

The use of MOx is vigorously opposed by some commentators. The principal issues of concern are:

- the implications for proliferation and civil liberties of widespread shipments of plutonium-bearing materials from which the plutonium could be relatively easily extracted – this claim is disputed;
- the nature of MOx spent fuel, which has a significantly higher proportion of actinides such as curium-244, plutonium-238 and americium-241, produces more heat and requires a longer period of interim storage than conventional spent fuel;
- the more difficult reactor physics associated with use of MOx in comparison with uranium fuels, making reactor control less straightforward.

It seems likely that plutonium fuel could be used in HTGRs, though these reactors have not been demonstrated on a commercial scale. The other main approach to plutonium destruction would be through the fast reactor. As discussed earlier, FR programmes in some countries have been expensive and success has been uneven. The main reason for their abandonment, sometimes without ever being brought into use, has perhaps been the declining perception of impending shortages of uranium. Although FRs are more efficient burners of plutonium than thermal reactors, they can also be designed and used to 'breed' plutonium in a 'jacket' made from uranium-238, and this produces plutonium that is weapons-grade and has potential proliferation implications. However, the objections to shipments of plutonium raised by opponents of MOx fuel might apply with greater force to a world in which significant transportation of plutonium-bearing fast reactor fuel was carried out.

Even without the use of plutonium-bearing fuels, an increase in thermal nuclear generation based on uranium might have proliferation implications, simply by increasing the amount of processed, unirradiated uranium in circulation – a material much easier to handle and enrich for producing weapons-grade material than spent fuel.

Furthermore, research or materials reactors might technically be used for proliferation purposes. Indeed, the requirement for isotopic separation, when producing radioactive materials for medical purposes for example, might make this more attractive to potential proliferators than civil nuclear power. In view of the vital role these facilities play in nuclear medicine, non-destructive testing etc., it would seem unlikely that they would be phased out even if nuclear power were to decline. They can also be built quickly – the Hanford production reactor was constructed in four months.

It would seem to follow that application of IAEA-type safeguards will be necessary for smaller reactors, even if large power reactors are phased out in the developed world. The issues of how to train and fund a body of inspectors without imposing enormous costs on research and production reactors would also need to be addressed.

Sidestepping the problem? The thorium cycle

From time to time it is suggested that a nuclear programme based on thorium rather than on uranium or plutonium might have advantages in terms of waste management and/or prevention of proliferation. Research has been under way for some decades, especially in countries such as India with large thorium deposits. Thorium-232 is not itself fissile, but it captures neutrons and is in the process converted into uranium-233, which is fissile. A mixture of thorium-232 with a source of neutrons such as uranium-235 or plutonium-239 can therefore be used as a reactor fuel. Alternatively, thorium-232 can be used in sub-critical assemblies.

Advocates of the process suggest a number of advantages. The spent fuel contains about one-third less transuranic elements than spent fuel from the uranium cycle. Thorium dioxide has a very high melting point (3,300°C, some 500°C higher than uranium dioxide), allowing the reactor to be operated at higher burn-up levels. (High burn-up produces higher levels of fission products and thus more decay heat, which thorium oxide fuel would be better able to withstand.) This also makes the fuel attractive for use in high-temperature gas-cooled reactors. The plutonium present in the fuel is burned up in the process, offering another way of destroying plutonium through utilization. The radioactivity from mine tailings from thorium decays more rapidly than that from mining uranium.

However, there are drawbacks as well. Uranium-233 is a rather attractive material for making nuclear weapons. Spent fuel from the thorium cycle includes thallium-208 and uranium-232, which emit extremely penetrating gamma radiation. Uranium-233-based fuels must therefore be handled remotely in gamma-shielded environments, an expensive technique

In conclusion, although the thorium cycle has attractions, especially in countries with significant thorium reserves, it remains to be proven technically and economically, and in any case it cannot be described as proliferation-proof. It remains very likely that the majority of future nuclear power stations will run on uranium and/or plutonium.

Summary

Waste management

It is a commonly expressed view in a number of countries that progress in radioactive waste management – or at least the management of radioactive waste from nuclear power operations, the same argument rarely being applied to medical and other non-power industrial wastes – is a prerequisite to further nuclear development. At the same time, there is a widespread view on all sides that there is a moral obligation on those currently using nuclear power to identify existing wastes, to create an infrastructure capable of implementing a long-term management strategy and to make financial provision for implementation. The advantage of relatively small volumes of the waste, by industrial standards, are countered by the very high and, in some cases, uncertain unit costs of disposal of some of the material, especially if it is assumed that there is no threshold below which radiation is harmless.

The claims of the nuclear industry that waste management presents no particularly difficult technical issues are hotly disputed by the industry's opponents. Moreover, the industry generally argues for deep disposal of radioactive wastes, both because this approach allows the insertion of several barriers of different kinds between the waste and the surface and because it would relieve future generations of the responsibility of managing the material. The terrorist attacks of 11 September 2001 seem to have strengthened this argument. But the industry's opponents argue that the current state of knowledge, both of basic science and of the characterization of particular sites, is insufficient to guarantee that material would not leak from a deep disposal facility before its activity had decayed to acceptable levels. Sometimes it is claimed that it will never be possible in principle to demonstrate the safety of disposal facilities over very long periods of time, so continuous active surface or underground monitoring may be a better approach.

Whatever the relative merits of these claims, it seems clear that given the failure to make progress in establishing waste disposal sites – or even appropriate decision-making procedures – in many countries, a period of on- or near-surface storage, possibly at the site of origin, will be necessary. Designing these stores, and the packaging of the wastes, for periods of time measured in decades or perhaps a century or more, would seem to offer a degree of flexibility to future decision-makers. It would allow for deep disposal if and when a robust safety case can be demonstrated but would create a suitable environment for storing the wastes in the interim. However, for many people the immediate future is more of a concern than the long term, and it cannot be assumed that people will accept this policy, even if they have shown little concern about the presence of the material on site in the past. Clear evidence that a long-term management approach had been identified and that there was the will to implement it might be necessary to achieve public acceptance.

In the longer term there may be other possibilities. Research into methods of partitioning the waste in order to separate out the longer-lived components and into transmuting these materials into shorter-lived radioisotopes is being pursued. Although commercial demonstration of these techniques is some distance away, the possibility has been raised of reducing significantly the amount of longer-lived materials requiring disposal. Transmutation may be possible in reactors (particularly fast reactors) or in other power systems such as sub-critical assemblies. However, there is dispute about what proportion of longer-lived fission products could be transmuted in this way and about the amounts and types of secondary waste (requiring management) that would arise in the course of transmuting certain radioisotopes. More development work on the P&T process will be required to settle such disputes. In any case, P&T would involve a trade-off between higher levels of radioactivity in the short term and reduced radioactivity in the longer term.

The possibility of developing international waste management facilities is also much discussed. It would seem unfeasible that each of the more than 60 countries operating nuclear power plants or research reactors and every country using radioactive materials for medical uses (most of the world) should have its own long-term waste storage facilities for all categories of waste, including the higher levels. Facilities for managing waste from countries with few nuclear facilities, which can use the collective expertise of the international community, would seem appropriate.

Whether they should be developed for the power wastes of major nuclear countries is more controversial. Although the pooling of resources and the choice of the most suitable sites from a number of countries would seem attractive, this policy would also require increased transportation of radioactive wastes, and might lessen the pressure on countries to reduce the volumes of waste arising from their domestic nuclear activities. Moreover, this policy could meet considerable political objection, especially from within potential host countries, and it would seem to cut across the ethical position that each country should be responsible for its own waste products.

Reprocessing

The case for reprocessing is likely to be affected significantly by perceptions of the future of nuclear power production. At present, a relatively small proportion of the spent fuel arising from nuclear power plants is reprocessed. Should the world nuclear industry decline or remain at approximately today's levels, there are unlikely to be major shortages of uranium for thermal power production for many decades. However, should the nuclear industry expand greatly, questions may arise about the economic and resource case for separating out for reuse the uranium and plutonium in spent fuel (which account for 97 per cent of the material). There may be questions too about the eventual requirements for spent fuel disposal facilities – clearly much smaller volumes of highly radioactive materials would require disposal if reprocessing were to be pursued. However, the requirements of venting heat production may mean that the size, and cost, of a high-level waste repository could be similar with or without reprocessing. (But the reprocessing route would also involve the production of quantities of intermediate- and low-level waste that would need to be managed.)

Reprocessing also has disadvantages. The Purex process produces significant volumes of waste, and discharges to the environment from reprocessing plants have historically been much higher than from operating nuclear stations, although the nuclear industry would argue that they still do not represent a significant environmental hazard. The production of highly active liquid waste as an intermediate between the spent fuel and immobilized waste is also potentially hazardous. The separation of plutonium represents a potential proliferation risk, and is also expensive. Although new approaches to reprocessing, such as pyrochemical methods, may produce fewer discharges and less waste, the other issues would still require resolution.

Proliferation

Nuclear weapons predate civil nuclear power – the time between the first demonstration of a nuclear chain reaction and the first atomic weapon test was barely two and a half years with 1940s technology, albeit under wartime conditions. It is extremely unlikely that a determined state, or perhaps even a sub-state organization, could be prevented from developing a uranium-based weapon, which would not even require reactor facilities. The existence of a large world nuclear industry could aid the proliferation of nuclear weapons, both by increasing the amount of fissile material in circulation and by allowing access to a body of nuclear expertise.

The offer of help with developing peaceful nuclear technology as a bargaining tool with which to prevent countries from diverting materials to military uses has been the basis of the Nuclear Non-Proliferation treaty for over 30 years. This treaty has largely been successful, although some countries have not become signatories and have developed their own weapons capability, while the nuclear weapons states have not disarmed (as they are required to do under the terms of the NNPT).

If plutonium remains mixed with highly radioactive fission products, it is in effect immobilized and made extremely difficult to use in weapons manufacture. However, the radioactivity of the fission products decays relatively rapidly, and within centuries spent fuel might become an attractive source of plutonium. It can thus be argued that for discouraging long-term proliferation, ways should be sought to destroy the plutonium rather than to immobilize it. These methods might include using it for power, in the form of MOx, in an inert matrix or in fast reactors (without breeding), or transmuting it in subcritical assemblies or in other kinds of reactor. However, none of these methods can be 100 per cent effective. Should nuclear power continue or expand, more proliferation-resistant fuel cycles may well have to be considered.

References

Barker, F. and M. Sadnicki (2001), *The Disposition of Civil Plutonium in the United Kingdom*, *fbarker@gn.apc.org*.

BNFL (2001), contribution to RIIA project on nuclear power (personal communication), *david.mc.horsley@bnfl.com*.

European Commission (1997), *ExternE Core Project Phase III Report*. Brussels: EC.

European Commission (2000), *Green Paper – Towards a European Strategy for the Security of Energy Supply*, COM(2000) 769. Brussels: EC.

Finch, C. (2001), Nirex (UK), personal communication, *Chris.Finch@nirex.co.uk*.

Häfele, W. (1998), 'Nuclear energy and waste: a global perspective', second Gordon research conference on nuclear waste and energy, 2–6 August 1998, Salve Regina, RI, United States.

House of Lords (2001), *Managing Radioactive Waste: Report of House of Lords Select Committee on Science and Technology*, Appendix 2. London: The Stationery Office.

IAEA (1974), *IAEA Yearbook 1974*. Vienna: IAEA.

IAEA (1997), *International Symposium on Nuclear Fuel Cycle and Reactor Strategies: Adjusting to New Realities*. Geneva: IAEA.

IAEA (2001), Research reactor database (RRDB). Vienna: IAEA.

Knight, J. L. (1998), 'Use of natural analogues in waste disposal', *Interdisciplinary Science Review – a special issue on radioactive waste*, 23 (3), September.

Loux, R. (1998), *Yucca Mountain: Follow the Money*. Carson, Nevada: Nevada Governor's Office Agency for Nuclear Projects, *www.state.nv.us/nucwaste/yucca/loux05.htm*.

Makhijani, A. (2001), 'Plutonium end game: stop reprocessing, start immobilising', *Science for Democratic Action*, 9 (2), February.

NEA (1993), *The Cost of High-level Waste Disposal in Geological Repositories*. Paris: NEA/OECD.

Royal Commission on Environmental Pollution (2000), *Energy: The Changing Climate*, CM4749, London: The Stationery Office.

RWMAC (2001), *RWMAC Advice to Ministers on the Radioactive Waste Implications of Reprocessing*. London: DETR.

Schneider, M., X. Coeytaux, Y. Faïd, I. Fairlie, D. Lowry, Y. Marignac, E. Rouy, D. Sumner and G. Thompson (2001), *Possible Toxic Effects from the Nuclear Reprocessing Plants at Sellafield (United Kingdom) and Cap de la Hague (France)*. Paris: WISE.

Smellie, J. and F. Karlsson (1996), *A Reappraisal of Some Cigar Lake Issues of Importance to Performance Assessment*, SKB Technical report 96-08, Sweden.

Strupczewski, A. (1999), 'Comparative assessments of emissions from energy systems', *IAEA Bulletin*, January.

Uranium Institute (1996), *The Recycling of Fissile Nuclear Materials: Final Report of the Working Group*. London: Uranium Institute.

US Department of Energy (1997), *Non-proliferation and Arms Control Assessment of Weapons-usable Fissile Material Storage and Excess Plutonium Disposition*, DOE/NN-007. Washington, DC: USDOE.

Venneri, F., N. Li, M. Williamson, M. Hours and G. Lawrence (1998), *Disposition of Nuclear Waste Using Subcritical Accelerator-driven Systems: Technology Choices and Implementation Scenario*, LA-UR 98-985, Los Alamos National Laboratory, United States.

WISE (2001), *Possible Toxic Effects from the Nuclear Reprocessing Plants at Sellafield (United Kingdom) and Cap de la Hague (France)*. Paris: WISE. *www.wise-paris.org/english/reports/STOAFinalStudyEN.pdf.*

5 Nuclear safety

Introduction

The safety of nuclear installations is a key issue, both in its own right and as a factor in public perceptions of nuclear technology. This chapter discusses some of the difficulties inherent in assessing and managing nuclear safety issues. It also reviews various approaches to ensuring the safe operation of present and future nuclear power stations, and other nuclear facilities, under conditions of major growth, modest growth and phase-out.

Inevitably, the focus will be on low-probability, high-consequence risks and the particular difficulties they cause for public policy-makers. The consequences of a major nuclear accident, say on the scale of Chernobyl, would be higher than the consequences of a single event associated with any other energy source, with the possible exception of large-scale hydropower. However, it should not be forgotten that the cumulative effects of a large number of low-consequence events, be they routine or accidental, could be as serious, or more so. Emissions of greenhouse gases from the use of fossil fuels and the cumulative effect of a large number of motor vehicle accidents are obvious examples from within the energy sector.

The nuclear power industry has long asserted that it works to safety standards far in excess of those for other industries. It is claimed, for example, that the assumed 'value of a human life' underlying the technology for preventing nuclear accidents (and the routine emission of radioactive materials) is higher than the equivalent figure used elsewhere; that, in other words, resources spent on improving nuclear safety could have better results if spent elsewhere. However, others hold that the consequences of a major nuclear accident would be so catastrophic that all technically achievable safety measures should be taken, irrespective of cost.

This is a complex argument. But whatever the merits of the two viewpoints, it is likely that the stringency with which nuclear safety is approached will only increase. First, as discussed later, there is the inevitable uncertainty, both about the likelihood of a major release of radioactive material (caused either by plant malfunction or by external events such as terrorist attacks) and about the consequences of such a release in any individual case. An application of the precautionary principle would thus imply that all reasonable steps should be taken to prevent releases, within a context of cost-benefit analysis.

Second, the perception of risks from radiation is not straightforward. Radiation has a number of features that seem to result in the perception of it as especially dangerous, an issue covered in more detail in Chapter 2, on public perceptions. In particular:

* radiation is seen as a risk that is relatively new, unfamiliar and undetectable by unaided human senses;
* radiation is regarded as being outside the control of the individual running the risk;
* radiation represents a (very small) threat to a very large number of individuals, including future generations, rather than a (larger) threat to a smaller number of people (such as coal miners or victims of a hydropower dam failure).

In addition, the nuclear industry is perceived by some sections of society as secretive and arrogant, and its pronouncements are treated with a degree of scepticism.

As a result, the fears caused by nuclear power radiation tend to prompt a particularly robust political response. It seems entirely legitimate that regulators should take into account, for example, people's unease with high-consequence, low-probability risks. However, it may also be argued that there is a danger of increasing the perceptions of the risks involved should the response be too robust.

It is generally recognized that the safety record of nuclear power stations has been good in the near half-century since the first commercial reactors opened. This record has been maintained despite particular concerns about the safety of reactors in the former communist countries of eastern and central Europe. With the exception of the criticality incident in Tokai-Mura, Japan in September 1999, which cost the lives of two workers in a facility fabricating research reactor fuel, there have been no radiation-related fatalities in nuclear installations since the Chernobyl accident in the Ukraine in 1986. This accident, which occurred in the design known as RBMK, which was unique to the former Soviet Union, remains the only civil nuclear reactor accident with demonstrable off-site health consequences (notably, over 2,000 cases of thyroid cancer) in nearly 10,000 cumulative reactor-years of commercial operation. The other major event, in a pressurized water reactor (PWR) at Three Mile Island, Pennsylvania in 1979, did not result in significant releases of radioactive materials, although some commentators claim there was an element of luck involved. (In addition there have been accidents in research reactors, some of which have led to fatalities among plant personnel.) An analysis of these two accidents is important, if only because they are the only large-scale 'case studies' available.

These observations raise questions along the lines of 'How safe is safe?'. If the current plants in use in most areas of the world have never been responsible for an accident with off-site consequences, in what sense are they 'unsafe', and what level of 'safety' would be 'acceptable'? These are not questions that can easily be answered by reference to reactor engineering and design.

Bodies such as the Health and Safety Executive (HSE) in the United Kingdom have examined the tolerability of various societal risks, taking into account not only physical harm but also factors such as ethical and social considerations.[1] It classifies risks as:

- 'intolerable' – they cannot be justified except in extreme circumstances;
- 'broadly acceptable' – they are insignificant, and attempts to reduce them further would be likely to result in a waste of resources;
- 'tolerable' (between these extremes) – measures should be introduced to drive these risks down towards the 'broadly acceptable' region.

In the 'tolerable' region, a particular degree of risk is regarded as acceptable if risk reduction is impracticable (the ALARP principle: as low as reasonably practicable) or if the cost of reducing it is grossly disproportionate to the reduction in risk achieved. The HSE found that an annual risk of death of about one in a million delineated the boundary between the 'broadly acceptable' and the 'tolerable' regions and that the boundary between 'tolerable' and 'intolerable' was about one in 1,000 for workers or one in 10,000 for members of the public. The concept of a limit of tolerability has been translated into basic safety limits (BSLs), and any new nuclear facility must satisfy these limits in order to be considered for licensing.

The HSE has emphasized that quantitative criteria are only part of the story in managing risks. Other important factors are understanding and presenting uncertainty, meeting fundamental engineering criteria and taking account of public perceptions of risk. According to the HSE's definitions, the risks associated with modern nuclear reactor designs would be expected to lie in the 'tolerable', if not 'broadly acceptable', regions.

Nevertheless, nuclear power retains the potential for catastrophic failure, on a single-incident scale far greater than that of any other energy technology

---

[1] HSE (1992).

with the possible exception, as noted, of large-scale hydropower. Assuming that the World Health Organization's appraisal of the Chernobyl accident (as supported by the UN Scientific Committee on the Effects of Atomic Radiation[2]) is correct – that demonstrable health effects have been restricted to people on the site at the time of the accident and immediately afterwards and to the 2,000-plus cases of thyroid cancer in Belarus, Ukraine and Russia – the accident ranks among the most serious in the history of technology. The financial cost of evacuation, decontamination etc. has been enormous. The human cost of evacuating peasant populations from their ancestral homelands and dispersing them in cities throughout the former Soviet Union is high. Significant areas of land will be contaminated for many years, with potential radiation-related health problems and costs in financial and psychological terms. Some commentators argue that the official estimates seriously underestimate the health effects of low levels of radiation, and claim that rumours persist of significant health problems in the region, in addition to those officially recognized. (Other commentators argue to the contrary, that the official risk estimates associated with very low levels of radiation are exaggerated and therefore that the long-term harm to human health caused by Chernobyl will be very limited.)

A central role in coordinating the safety of nuclear installations worldwide has been played by the International Atomic Energy Agency, which was formed in 1957 to serve as the world's intergovernmental forum for scientific and technical cooperation in the peaceful utilization of nuclear energy. Its principal objectives under its statute are 'to accelerate and enlarge the contribution of atomic energy to peace, health and prosperity throughout the world' and to 'ensure, so far as it is able, that assistance provided by it or at its request or under its supervision or control is not used in such a way as to further any military purpose'. (Some commentators have pointed to the possible conflict of interest between the objectives of promoting nuclear technology and being a quasi-regulator.) These objectives are pursued by promoting the transfer of nuclear technology and know-how, encouraging the creation of an international culture of safety and reliability in the utilization of nuclear energy, safeguarding nuclear materials so as to ensure that they are used exclusively for peaceful purposes and disseminating information on the peaceful uses of nuclear technology. Over 130 countries are members of the IAEA.

A further point is important. It is a feature of technological evolution that particular technologies tend to become safer over time. The current focus, for

---

[2] UNSCEAR (1988).

example on passive safety features, arises in part from a desire to reduce the likelihood of major reactor core incidents. However, those considering purchasing these designs will often also operate existing stations of an older design. There is a tension between arguing that new reactor designs are safer but that existing designs are safe enough to continue operating. This tension exists in all technologies, but is perhaps particularly salient in the nuclear case. The projected lifetime of a modern nuclear reactor is about fifty years, during which time a number of technical discoveries might be expected. Moreover, existing stations have proved to be 'safe', in the sense that there have been no major accidents with demonstrable off-site health consequences in any reactor design; but the potential consequences of a major accident are enormous. In part the likelihood of major accidents can be reduced by backfitting safety devices and upgrades in accord with the general technical standards that are necessary for existing plants to gain regulatory permission for continued operation and lifetime extension.

Particular issues in nuclear safety

It is difficult to develop a realistic way of comparing technologies characterized by high-frequency, low-consequence accidents with those which have low-frequency, high-consequence effects (and which may, of course, have a very large cumulative effect). First, it is unlikely that all possible scenarios – all possible failures of mechanical, electrical or human elements – that could lead to a major accident can ever be identified with complete certainty. In particular, the behaviour of the plant in a severe accident may be very different from its behaviour under the normal conditions for which it had been designed. For example, although instrumentation and display systems should be designed to cover credible accident scenarios, the experience at Three Mile Island showed that overload, both of the instrumentation and the human operators, may cause severe difficulties. At one stage, the printer downloading alarm messages was receiving over 100 messages per minute, well beyond its operational capability. Eventually the backlog of information at the printer exceeded 30 minutes, and operators had to 'dump' large amounts of it.[3] Similarly, it is unlikely that all of the possible ways that might be open to determined terrorists to cause damage at nuclear installations could be identified with any certainty.

    Second, although light water reactors dominate global nuclear power production, most reactors differ from one other in the details of design. This has

---

[3] Kemeny et al. (1981).

prevented the compilation of a statistical database for component and system reliability to which designers can refer. This is in contrast with, for example, the aircraft industry, where very large numbers of identical aircraft have been produced and thus meaningful performance databases have been constructed.

Third, there is considerable difficulty in assessing or predicting the damage caused by a major accident. As noted on page 150, there is still considerable dispute over the actual consequences of the Chernobyl accident. The direction of the wind and the presence or absence of rain in the immediate aftermath of a major release of radioactive materials can have a profound influence over the deposition of those materials. The number of people living near the plant and the effectiveness of local emergency responses and health care will also be relevant. (This is of course applicable not only to nuclear technology. The earthquakes in Armenia in December 1988 and in California in January 1994, comparable in physical severity, led to very different levels of fatalities – between 25,000 and 100,000 in Armenia and 56 in California – demonstrating different levels of structural safety.)

An uncomfortable situation arises from these observations. It is generally accepted that a large nuclear accident would have consequences that were unacceptable, or bordering on the unacceptable, but such an accident cannot be ruled out, at least in view of the technology used in the majority of today's plants and perhaps not in principle.

In addition, there is uncertainty as to the effects of very low radiation doses. There is little dispute that levels of radioactivity released into the environment from routine operations at nuclear installations are small compared to background levels. Doses caused by the civil nuclear industry to most of its workers, and to practically all people living offsite, are therefore much smaller than doses received because of natural radioactivity. Further, of the doses caused by the activities of man, by far the largest are associated with the medical uses of radiation. However, there is some debate as to whether natural and artificial radioactive materials have similar biological effects. It is clear that human tissue is unable to distinguish between identical radiation doses from natural or man-made isotopes, but the tendency of some artificial isotopes, such as caesium-137, strontium-90 and plutonium, to accumulate in human tissue is a complicating factor. There is considerable confidence among bodies such as the International Commission on Radiological Protection that these considerations are well understood and are taken into account in current models. It is also argued that the effects of radiation have been studied more than those of any other type of hazard. A series of studies has failed to find health problems clearly associated with nuclear power, although there have been higher than expected levels of leukaemia near some older nuclear establishments, notably Sellafield in the United Kingdom and Cap de la Hague in France. Some sceptics point to the fact that reappraisals of the calculated damage caused by doses of radiation have each concluded that radiation was more dangerous than previously thought.

There is much certainty about the short-term effects of high doses, but views about the effects of radiation at low dose rates cover a wide range. At one end of the spectrum is a belief that radiation is more dangerous, dose for dose, at low dose rates than at high ones, owing to genetic effects that are non-fatal at a cellular level. Some support the 'linear no-threshold hypothesis', which holds that any exposure to radiation carries a proportional risk. Next comes a belief that low levels of radiation have no effect one way or the other. Finally, there are those who hold that below a certain dose rate, radiation becomes beneficial as it stimulates the immune system. This view, they claim, has much supporting evidence, although it has tended to be dismissed by regulatory authorities who, on the basis of the precautionary principle, make use of the linear no-threshold principle.

Much of the current understanding of the health effects of low doses of radiation has been developed from a long-term study of the survivors of Hiroshima and Nagasaki. The most important dose sustained by this population came from outside the body and was instantaneous. There may be differences between this kind of dose and longer-term doses originating from radioactive materials within the body, although both animal studies and studies on other human populations suggest that different types of dose are broadly comparable.[4] However, the question is not settled.

It is very unlikely that statistics alone will have sufficient power to settle conclusively which of the various views about the health effects of low-level radiation is the correct one; the effect, if any, is too small to be distinguished from random fluctuations in cancer rates. As a result, increasing attention is being paid to the underlying cell biology. In the meantime, the linear no-threshold hypothesis has been adopted by international regulatory bodies as a guide to policy for radiological protection. This accords with the precautionary principle that where uncertainty exists over environmental discharges, it should be assumed that they are harmful until there is evidence to the contrary.

The approach to radiological protection concerning routine operations has three strands:

---

[4] UNSCEAR (1988).

- justification – no dose can be sustained without the activity involved offering a clear and sufficient benefit;
- ALARP – all doses must be as low as reasonably practicable;
- dose limits – no dose can be sustained which breaches nationally set limits.

This regime clearly has little relevance to emergency situations, however, except in protecting emergency workers from receiving unacceptably high doses.

A further aspect of nuclear safety, distinct from concerns about design flaws, concerns the fate of reactors in politically or technologically unstable regions or times. Unlike many energy technologies, nuclear reactors, especially if the spent fuel has not been removed, represent a long-term health, environmental and security risk. The decommissioned Soviet submarines in the Barents Sea are a notable example, and it is believed that a number of reactors in the former Soviet Union pose similar risks. A related issue, which gained extra prominence following the events of 11 September 2001, is the potential vulnerability of nuclear installations to suicide terrorist attacks. The danger could be twofold – either the risk of radioactive releases or the risk of having to close down a number of large power plants indefinitely as a precautionary measure, with threats to secure supplies of electricity.

Principles of reactor safety

The safety of a nuclear reactor depends on its ability, under all circumstances, to prevent significant amounts of radioactive materials escaping into the environment.

There are perhaps three primary safety imperatives that a reactor must obey:

- to keep the core power down, by controlling the power level and stopping the fission chain reaction when necessary;
- to keep the core cool, by maintaining adequate cooling to prevent core meltdown owing to decay heat in fission products after the reactor has been shut down;
- to keep radioactive material contained as close as possible to the source, under all conditions.

Three approaches, which are not mutually exclusive, can be taken to achieve these imperatives. They are the employment of passive safety systems, the employment of active (or engineered) safety systems and the development of an appropriate safety culture among operators. Passive safety relies on forces of nature such as changes in density at certain temperatures or gravity, while engineered safety relies more on technical features such as valves or pumps.

Broadly speaking, it is easier to use passive safety features in smaller reactors than in larger ones. For example, the surface-area-to-volume ratio for a smaller reactor is greater than for a larger reactor of the same or similar type. As heat production is proportional to volume and heat dissipation is proportional to (surface) area, the ratio of passive cooling to heat production is higher in smaller reactors. This means that proportionally less cooling water is required in emergency scenarios, and it is easier to provide this water without requiring pumps. But now passive safety and simplified reactor design are being applied to new designs of larger reactors as well.

A related issue is the physics of the reactor itself. In the pressurized water reactor (unlike the Soviet-designed RBMK), the water coolant serves as the main moderator as well as for cooling. If the coolant escapes, nuclear fission stops (because a moderator is necessary for fission to occur in a thermal reactor). There could still be an extremely serious accident, as waste decay heat in the reactor fuel would still need to be dissipated, but there is much less likelihood of a runaway surge of power in a very short time. It is argued by some commentators that there is less need for multiple redundancy in passive safety systems (see page 156), as this kind of safety does not rely on particular pieces of machinery operating as designed. It is further argued that the behaviour of passively safe systems in emergency situations is easier to predict with certainty.

Engineered safety systems might include mechanical shutdown apparatus and emergency cooling equipment. These systems depend on machinery doing the job for which it was designed rather than on the passive use of the laws of physics. In addition, the complexity of operation of a large modern nuclear power station, results in considerable reliance on automatic control and surveillance. Advanced automated systems are especially important in the early stages of a potential accident in diagnosing the cause of malfunctions, assessing their possible evolution and taking appropriate corrective action.

The third element in safe operation requires that personnel are trained and directed to operate the technology within certain defined limits. This is especially important if the plant is engineered to operate in a way in which, without any component failure, major accidents are possible should operators fail to follow the prescribed procedure. Perhaps the most obvious example of this kind of engineering design was the RBMK, the Soviet-designed reactor involved in the Chernobyl accident. The operation of the plant at very low power could, and did, result in an uncontrollable power surge. (Although not relevant to reactor safety, it might be noted that at Tokai-Mura in 1999, the design of the steel holding tanks was such that a critical quantity of enriched uranium could be introduced.)

All else being equal, the first of these three approaches is the most attractive, as it is not susceptible either to component or human failure. However, it is not clear that sufficient passive safety features can be incorporated in a reactor design so as to respond to all accident scenarios, especially for big reactors with a correspondingly lower surface-area-to-volume ratio. Most of the nuclear stations operating today therefore also rely at least to some extent on a combination of the second and third approaches.

The ideal of nuclear safety design is to develop a 'forgiving' reactor system in which no actions by the operators can result in a major release of activity. At both Chernobyl and Three Mile Island, actions by the operators, whether carried out in error or in deliberate contravention of operating procedures, caused or exacerbated major incidents. That there was no significant release of radioactivity at Three Mile Island, despite the fact that 3,000 MW of thermal power was running out of control, suggests that forgiving technology may be achievable, at least to the extent of containing any credible accident. It is unwise, however, to draw too many conclusions from a single major incident.

Defence in depth and multiple redundancy

Two key elements of the safety philosophy in Western reactors have been 'defence in depth' and 'multiple redundancy'. The fundamental principle of defence in depth is that should one safety level fail, the next level will automatically come into play. Its importance as a fundamental strategy for achieving safety has been reaffirmed several times. The IAEA has suggested a more structured interpretation of the concept, and has focused on future reactors.[5]

The defence-in-depth concept generally applies on five levels:

- prevention of abnormal operation and failures;
- control of abnormal operation and detection of failures;
- control of accidents within the design basis;
- control of severe plant conditions, including prevention of accident progression and mitigation of the consequences of severe accidents;
- mitigation of radiological consequences of significant releases of radioactive materials.

---

[5] IAEA (1996).

The objective of the first level of defence is to prevent abnormal operation and system failures. If this level fails, failures are detected and abnormal operation is controlled at the second level of defence. Should the second level fail, the third level ensures that safety measures are carried out to prevent core damage by activating specific engineered safety systems. Should the third level fail, the fourth level limits further accident progression by way of design features and accident management procedures, developed to prevent or mitigate severe accident conditions. The last objective, at the fifth level of defence, is mitigation of the radiological consequences of significant off-site releases through the off-site emergency response.

To take containment of fuel within a reactor as an example, a series of physical barriers are placed between the radioactive reactor core and the environment. The fuel generally consists of pellets made of metal or ceramic with a high melting point. Radioactive fission products remain bound inside the fuel pellets. In most Western reactors, the pellets are packed inside zirconium alloy tubes. The rods are contained inside a large steel pressure vessel with walls about 20 cm thick. This in turn is enclosed inside a concrete outer containment structure with walls typically a metre or more thick. There are control rods to absorb neutrons and cooling systems to remove excess heat.

Where possible, safety features such as auxiliary pumps for providing cooling water in the case of an accident are furnished with back-up in order to compensate for component failure or human error. Safety systems typically account for up to 60 per cent of the capital cost of large Western reactors of current design.

Major reactor accidents

Although very different in their consequences, the accidents at Three Mile Island in 1979 and Chernobyl in 1986 stand as the two major safety failures in the history of civil nuclear energy. A consideration of their course is thus of interest.[6]

A brief description of the accidents

At Chernobyl, operators were carrying out an experiment to check whether, in the case of a power cut affecting safety systems, there would be sufficient power in the turbogenerators as they coasted down for them to run emergency

---

[6] See Grimston (1997).

cooling systems until back-up power could be provided. In order to do this the operators ran the plant at very low power, a forbidden regime in which it was known that a runaway surge of power was highly likely. At Three Mile Island a relatively minor component failure – a relief valve sticking open – resulted in nearly half of the fuel in the core melting. Although there was no significant release of radioactivity, it took several hours to bring the plant under control, and it had to be completely written off.

In both accidents, technical factors were involved. At Chernobyl, the problem was not component failure – there was none. There were, however, design weaknesses, presumably regarded as tolerable by the designers. First, those weaknesses allowed the reactor to suffer a runaway surge in power when operating at low power (the 'positive void coefficient'). Second, they allowed operators to disable the emergency core cooling system, which would have prevented operation at low power. At Three Mile Island the relatively minor component failure that started the accident was exacerbated by the operators intervening in mechanized safety systems. Things were made worse by the design of some of the instrumentation. For example, the indicator showing the temperature in the 'reactor coolant drain tank', which might indirectly have alerted the operators to the valve problem, was located on the back of the main control panel, out of the sight of the operators.

Neither accident was characterized by human 'error', in the sense that the operators accidentally deviated from well-established practice. As noted earlier, at Three Mile Island the welter of confusing information coming to the operators made it practically impossible for them to determine the causes of the problem. As a result, their actions were not appropriate to the situation, in particular their overriding of the emergency cooling regime, which would have brought the situation under control. The Chernobyl accident was characterized by the operators seeming to operate the reactor quite deliberately in a way in which, in the words of Valery Legasov, head of the Soviet delegation to the conference on the accident held in August 1986, 'no one in the whole world, including the President of the country, could allow'.[7]

Organizational failures

An intermediate class of errors can be identified that is 'organizational' in character. Organizational inadequacies might involve excess pressure on operators to ignore safety procedures (or indeed inadequate safety procedures in the

---

[7] Ignatenko et al. (1989).

first place); poor management or other factors leading to low morale; poor supervision of operations; and a poor flow of information through the operating utility. These factors can result in operators quite deliberately acting contrary to safety and other operating codes or being unaware of them. (This collection of managerial and operational attitudes is often called the 'safety culture', a term that stresses the need for all concerned to regard safety as the highest priority.) In cases of organizational shortcomings, the term human 'error' would not seem to be appropriate.

Organizational failures leading to major accidents can be classified into five groups:

• an overriding production imperative – institutional pressure to maintain production and also to 'get the job done on time';
• failure to allocate adequate or appropriate resources;
• failure to acknowledge or recognize an unsatisfactory or deteriorating safety situation;
• lack of appreciation of the technical safety envelope;
• failure to define and/or assign responsibility for safety.

At Chernobyl, the experiment was designed by a consulting electrical engineer, who assumed that the reactor would have been shut down by the time the experiment was to be carried out, and thus did not consider the effect of the experiment on its operating limits. The operators kept the plant running at low power, presumably because they wished to be able to run the experiment a second time should the first attempt be inconclusive. Members of the state inspectorate, the Gosatomenergoadzor, had all gone to the local clinic for medical inspections on 25 April, so nobody was on site to prevent breaches of the operating code. Also, a last-minute demand from the local electricity control centre resulted in the plant being run at half-power for nine hours, significantly altering reactor fuel characteristics. The test had been scheduled for the afternoon of 25 April, but it was carried out in the early hours of the next morning, when most of the site's scientists and engineers had left, and perhaps also when the operators were not at the peak of alertness. (It is notable that the chain of events leading to the accidents at Chernobyl, Three Mile Island and Bhopal (see page 161) started in the early hours of the morning.) In carrying out the test, the operators overrode the many 'trip' commands coming from various parts of the plant. Indeed, the whole process of keeping the reactor at power, in order to repeat the experiment if necessary, was not part of the test programme; the test could have been carried out soon after switching the reactor off, using the decay heat of the fuel.

A sequence of events similar to the early stages of the Chernobyl accident had occurred at the Leningrad (St Petersburg) RBMK plant in 1982. The information about this does not seem to have reached the operators.

There seemed to be a remarkable complacency at Chernobyl about safety, perhaps because the nuclear industry was seen as a flagship of Soviet technology. One site manager was quoted as saying 'What are you worrying about? A nuclear reactor is only a samovar.' Operators were selected not only for their technical ability but also for their loyalty to the Communist Party. Officially, operating procedures, derived more from the plant design than from operating experience, were to be adhered to 'by the book'. Overtly to do otherwise would be to invite instant dismissal and a return to the 25-year waiting list for an apartment. In reality, however, operators were constantly being put into situations that conflicted with this imperative – for example the local mayor, a high Party official, demanding extra power during a cold spell, something that would be done if possible, whatever 'the book' said. Thus the highly talented workforce was daily discouraged from using personal initiative and taking responsibility for safety, but were quite used to bending the rules covertly.

As for Three Mile Island, perhaps the most obvious organizational issue was the failure to pass on to the operators the details of an incident in 1977 at the Davis Besse plant. That reactor, like the one at Three Mile Island, had been built by Babcock and Wilcox. A pressure-operated relief valve had stuck open, and operators responded to rises in pressurizer water levels by throttling water injection. Because the plant was at low power (nine per cent) and the valve closed after 20 minutes, no damage was done. However, an internal Babcock and Wilcox analysis concluded that if the plant had been operating at full power, 'it is quite possible, perhaps probable, that core uncovery [loss of cover of water] and possible fuel damage would have occurred'. In January 1978 a Nuclear Regulatory Commission (NRC) report concluded that if this circumstance had arisen, it was unlikely that the operators would have been able to analyse its causes and respond appropriately. But none of this information was passed on to other plant operators, either by Babcock and Wilcox or the NRC.[8]

There was also a lack of staff and expertise in the area of nuclear plant operation. For example, a review of technical information from other plants was carried out by people without nuclear backgrounds. In addition, the appor-

---

[8] Kemeny et al. (1981).

tionment of responsibility for safety both within the operating utility and within the regulatory authority was not clear.

The Bhopal accident in comparison

In some ways, the major accident at the Union Carbide chemical plant in Bhopal, India in December 1984 was similar to Chernobyl and Three Mile Island:

- operators disabling emergency cooling systems (against 'strict' instructions);
- runaway production of heat;
- no significant component failure or human 'error';
- poorly ascribed responsibility for safety;
- emergency starting in the early hours of the morning.

News of a similar incident in Union Carbide's plant in West Virginia was not passed on to local management at Bhopal. As at Three Mile Island, the Bhopal instrumentation was deficient – one critical dial, showing the pressure within one of the large storage tanks which exploded, was not even in the control room, and operating instructions were generally in English, which many of the operators did not understand.

At Bhopal, however, the morale of the operators was very low, unlike at Chernobyl, and the standard of training was poor in what was regarded as a dead-end plant in a loss-making division – indeed it was up for sale at the time of the accident. It would appear that morale that is either too high, leading to complacency, or too low can act as a threat to safe operation.

Lessons to be learnt

It is difficult to draw too many conclusions from a relatively small number of major accidents. However, the following points seem clear:

- a design approach should be taken that eliminates the possibility of runaway power surges;
- an engineering approach should be taken that places as few crucial decisions as possible in the hands of operators in the early stages of a potential accident;
- emergency safety systems should not be capable of disablement by operators or plant managers;

- there should be a culture of open discussion of any inadvertent deviations from operating instructions during normal operation and their consequences;
- every effort should be taken to make sure that plant operators are aware of relevant incidents at other facilities;
- a safety culture should be developed whereby operators are both aware of the potential consequences of a major deviation from safety procedures and confident that no pressure will be brought to bear on them to flout these procedures;
- repetitive and boring tasks require careful management supervision;
- in any scenario in which nuclear plants are not being replaced at the end of their lives, a comprehensive and consistent approach to the future of the operators should be regarded as integral to maintaining morale and therefore safety.

Many of these lessons have been incorporated into new reactor designs and operating codes, including requirements in many countries that no human action should be required within the first 30 minutes of an accident.

Measures taken since Chernobyl

Of the measures introduced in response to the Chernobyl accident, perhaps three are worthy of special mention. The IAEA Convention on Nuclear Safety entered into force in 1994, and has been accepted and welcomed by all sides of the international nuclear community. The convention obliges states to take national measures with respect to safety matters, including:

- legislative and regulatory frameworks;
- assessment and verification of safety, emergency preparedness and operation of nuclear power plants;
- reporting on measures taken to implement each of the obligations under the convention.

It thereby seeks to ensure that the use of nuclear energy is safe, well regulated and environmentally sound, and reaffirms the necessity of continuing to promote a high level of nuclear safety worldwide. It entails a commitment to the application of fundamental safety principles for nuclear installations, rather than of detailed safety standards. There are internationally formulated safety guidelines, which are updated from time to time and provide guidance on contemporary means of achieving a high level of safety.

The concept of 'safety culture' was introduced in 1991. Safety culture is the body of characteristics and attitudes which establish that, as an overriding priority, nuclear plant safety issues receive the attention warranted by their importance. Safety culture is a matter of attitudes as well as procedures; it relates both to organizations and individuals, and concerns the requirement to match all safety issues with appropriate attention and action. Safety culture is also an amalgamation of values, standards and norms of acceptable behaviour. These are aimed at maintaining a self-disciplined approach to the enhancement of safety beyond legislative and regulatory requirements. Thus, safety culture has to be deeply ingrained in the thoughts and actions of all individuals at every level in an organization. The leadership provided by top management is crucial.

A further development in the years since Chernobyl has been the adoption of the International Nuclear Event Scale (INES) (see Table 5.1). It was designed by an international group convened jointly by the International Atomic Energy Agency and the Nuclear Energy Agency (NEA) of the OECD, and reflects the experience gained from the use of similar scales in countries such as France and Japan.

Events are classified on the scale at seven levels. The lower levels (1–3) are termed incidents, and the upper levels (4–7) accidents. Events that have no safety significance are classified as level 0, or below scale, and are termed deviations. Each criterion is defined in detail in the *INES Users' Manual*. Events are considered in terms of three safety attributes or criteria: off-site impact, on-site impact, and degradation of defence-in-depth. If an event has characteristics represented by more than one criterion, its overall severity rating is that of the most serious characteristic. Chernobyl was classified as Level 7, Three Mile Island and the accident at the military reactor at Windscale in Britain in 1957, where a fire caused local releases of radioactivity, as Level 5.

Table 5.1 The International Nuclear Event Scale

Accident	7 Major accident
	6 Serious accident
	5 Accident with off-site risk
	4 Accident without significant off-site risk
Incident	3 Serious incident
	2 Incident
	1 Anomaly
Deviation	0 Below scale – no safety significance

The current situation

Perhaps the most pressing safety requirement at present, other than the need to assess possible terrorist threats to nuclear installations, is to ensure the safety of reactors in the former Communist countries of eastern and central Europe. However, issues exist in all areas.

The safety of facilities in the developed world

As discussed earlier, since Three Mile Island much effort has been directed towards plant development, in particular towards reducing the likelihood of operators interfering unhelpfully in automated safety systems. Following the Three Mile Island accident, the Institute of Nuclear Power Operators (INPO) was formed in the United States. Its mission is to promote the highest levels of safety and reliability in the operation of nuclear-powered plants generating electricity. All US organizations that operate commercial nuclear power plants have become INPO members, and participants now include operating organizations in other countries and also nuclear steam-supply system companies, architect–engineering companies and construction firms. The National Academy for Nuclear Training, which operates under the auspices of INPO, integrates the training-related efforts of nuclear utilities, the independent National Nuclear Accrediting Board and INPO's training activities.

Nonetheless, worrying incidents do occur from time to time. They suggest that constant vigilance is needed in order to maintain an appropriate safety culture. At Sellafield in the United Kingdom in mid-1999, it became clear that personnel responsible for carrying out quality assurance checks on mixed-oxide fuel pellets for export to Japan had been falsifying records. (In the aftermath of the accident at Chernobyl, workers there were found to have falsified safety checks.) Clearly, the management supervision of this extremely tedious and physically uncomfortable procedure was inadequate. Although there were no safety implications in this particular case, as the safety checks themselves were carried out mechanically, the incident did serve as a timely reminder that the safety culture cannot be taken for granted anywhere. It also undermined international confidence in BNFL's standards. The result was a root and branch re-examination of safety procedures at the plant.

At Tokai-Mura in Japan in September 1999, workers at a facility processing research reactor fuel poured solutions containing enriched uranium from steel buckets into containment tanks. A criticality incident occurred, leading to the deaths of two workers and to off-site contamination, but there were no detectable off-site health consequences. From the earliest days of nuclear technology, it was recognized that bringing too much fissile material together into a small space would result in a chain reaction, producing heat and radiation. Various methods have been devised to prevent unwanted criticality, but in this case the simplest of poor practices overcame them all.

These incidents demonstrate that weak safety culture is not the preserve of developing countries or economies in transition. It is a major challenge for managers in the industry to maintain a high level of vigilance in circumstances where no problem is anticipated, either because the task seems routine or because very long periods of time pass without any anomaly being identified.

The safety of reactors in the former Communist countries of Europe

After the accident at Chernobyl and the subsequent collapse of the economies of eastern and central Europe, there has been great concern about safety standards in the region. Quite apart from the damage that a major accident would do to local people and environments, a repeat of Chernobyl would, it is often observed, result in significant difficulties for the operation and construction prospects of nuclear stations in other countries, even to the extent of heralding the end of nuclear power.

Concerns about safety focus on two basic nuclear reactor designs developed by the former Soviet Union. One, the VVER, is similar in concept to the Western pressurized water reactor, and was widely exported to central and eastern Europe. The VVER was produced with two rated outputs. The earlier stations, consisting of two main series, each with a number of variations, have a rated output of 440 MWe; the later stations, also of two series, have a rated output of 1,000 MWe. The other, the RBMK, was built only in the USSR (Russia, Ukraine and Lithuania). RBMKs, of which there are three series, are mainly rated at 1,000 MWe, although the Lithuanian ones are rated at 1,300 MWe. The RBMK is also a light water-cooled design, but includes a (graphite) fixed moderator. Largely because of this feature, the RBMK has the potential to suffer runaway power surges when operating at low power. This operation is strictly forbidden, but was not, until recent modifications, physically impossible. This 'positive power coefficient' in one of the four RBMKs at Chernobyl was the main cause of the accident there. All four are now closed.

In early 2001, 63 reactors generated a total of some 45 GW of capacity in the former Soviet bloc and Finland (see Table 5.2)

In 1990 the International Atomic Energy Agency instituted a programme looking at safety issues affecting the earlier VVER-440s. It was expanded to look at all reactor types in central and eastern Europe. Its main aim was to

Table 5.2 Numbers of former Soviet reactors in operation, 2001

	VVER-440	VVER-1000	RBMK
Armenia	1	nil	nil
Bulgaria	4	2	nil
Czech Republic	4	1	nil
Finland	2	nil	nil
Hungary	4	nil	nil
Lithuania	nil	nil	2
Russia	6	7	11
Slovakia	6	nil	nil
Ukraine	2	11	nil
Total	29	21	13

develop books on safety issues for the main reactor types, listing and ranking relevant safety issues and making recommendations for improvements. The final report of the Programme on the Safety of VVER and RBMK Nuclear Power Plants was published in 1999.

A number of significant problems were identified for the earlier VVER-440 series, of which the most serious concerned the integrity of the primary cooling circuit and the quality of the confinement. No major issues were identified for the later VVER-440s or for the VVER-1000 series, suggesting that the development of safety features from the earlier VVER-440s had largely been successful. However, questions were raised about the quality and reliability of some components. The programme of work on the RBMK reactors has focused on remedying key design problems and on improving the reliability of some of the pipework. Modification work has been carried out to aid rapid shutdown, and consideration is being given to introducing additional shutdown systems.

Various factors of safety culture needing to be addressed were identified in a number of plants of all descriptions. These factors are of importance to all plant operators, and include:

- production functions predominating over safety considerations;
- a lack of openness and communication, insufficient dissemination of information, limited feedback on operating experience;
- an absence of exchanges with the international nuclear community, leading to stagnation in safety, coupled with poor monitoring of plant safety systems, personnel and procedures;
- insufficient maintenance of equipment, poor housekeeping;

- an absence of or unsuitable emergency operating procedures and also a tendency to train 'on the job', inappropriate for emergency situations;
- a complex organizational structure, with very little delegation and minor problems being taken to management level.

A further essential finding of the IAEA study was that standards varied widely, even among plants of the same series, owing to differences in the practices of owners, operators and national regulators, the level of modification of the basic Soviet designs and the availability of finance for improvements.

An important development in recent years has been a growing cross-fertilization of ideas and technology between the West and central and eastern Europe, as demonstrated, for example, by the use of Westinghouse controls and instrumentation in the Russian-designed VVER reactors at Temelín in the Czech Republic that were commissioned in 2000. (The VVERs at Loviisa, Finland, which began operation in the late 1970s, have Western containment buildings.) A number of other bodies, notably the OECD's Nuclear Energy Agency, the European Union and the European Bank for Reconstruction and Development, have been involved in promoting cooperation and assistance, especially in the areas of nuclear safety and nuclear legislation and liabilities. Considerable progress has been made. For instance, after Chernobyl a new IAEA code (OPB-88) introduced more stringent safety requirements for new designs. However, the conclusion of the final report of the IAEA VVER–RBMK Programme is stark: 'Despite the improvements in safety already achieved, much remains to be done at individual nuclear power plants, particularly at the VVER and RBMK plants of the first generation.'

In another initiative, the World Association of Nuclear Operators (WANO), set up along the lines established by INPO soon after the Chernobyl accident, has brought together the operators of nuclear stations from all around the globe. It has encouraged exchange visits, twinning arrangements, the exchange of information about incidents, the identification of 'good practice' and peer review of individual plants. WANO has been particularly successful, perhaps by dint of being not an organization of states but of operators. Its benefits have been felt not only in improved safety but also in improved operating performance.

A number of other initiatives have been instituted that involve bilateral aid or bodies such as the G-7, the G-24 and the European Union. However, positive results have sometimes been difficult to demonstrate.

The safety of installations in the developing world

In principle, the same safety issues apply in developing countries as in the developed world. Since Chernobyl, it has been as clear in developing nations, as elsewhere, that a major nuclear accident would have serious consequences for local people and for the prospects of further nuclear development. There is thus a considerable desire within countries such as China to ensure that nuclear development proceeds in accordance with international safety standards.

There are two broad ways in which this might be achieved – the use of international designs or the harmonization of national safety standards applying to indigenous nuclear industries. The purchase of nuclear designs, on licence, from construction companies in the developed world allows the host country to gain access to the operating experience of similar designs of plant. Several developing countries have started their programmes in this way. This may be helped by the consolidation within the nuclear construction industry, which has resulted in four or five major plant options being available now or awaiting demonstration.

However, there are also pressures within larger developing countries to establish an indigenous nuclear industry, both for economic and employment reasons. It is likely, then, that in the future more plants will have domestic elements, although some components, such as the pressure vessel, turbines etc., may still be purchased from overseas. In these circumstances, IAEA nuclear safety standards are increasingly important, as are links through other international organizations such as WANO. China, for instance, has used the IAEA standards, codes and guidelines as a reference in establishing domestic safety procedures.

It should be noted that sanctions, such as those imposed against India and Pakistan because of their atomic weapons tests in the late 1990s, may jeopardize collaboration between developed and developing countries over safety matters.

External threats

In addition to the risk of a major release of radiation caused by plant malfunction, whether for technical or human reasons, nuclear installations could in principle be breached from outside, either by terrorist action or because of a general breakdown in society. The terrorist attacks on the United States in September 2001 brought this possibility to the forefront, and resulted in a widespread assessment of the potential threats to nuclear plants from similar action.

Possible countermeasures include increased security measures on site, the use of anti-aircraft artillery or fighter aircraft and strengthened outer containment. However, it remains an open question as to whether a passenger aircraft, fully loaded with fuel, or some other projectile or weapon could be prevented from breaching safety systems, at least in older plants, if the pilot were prepared to die in the attempt.

The future safety of nuclear plants

In addition to the issues of safety culture and terrorism discussed above, there are three important factors that are likely to affect the future safety of nuclear plants:

* regulatory regimes, including those in place in developing countries;
* the concepts and designs of plants;
* the significance of the liberalization and commercialization of electricity supply systems.

Regulatory regimes

In general, there are four characteristics of good regulation:

* independence;
* a legal basis;
* technical knowledge;
* adequate funding

The potential global impact of civil nuclear power has been recognized since its very early days. Bodies such as the IAEA, EURATOM and the OECD's Nuclear Energy Agency were formed more than 40 years ago to manage this impact. Nonetheless, the history of nuclear power has to an extent been characterized by a series of national programmes with relatively limited cross-fertilization. Undoubtedly the genesis of the industry in the military programme has been a reason for this.

Also, the isolation of some parts of the global industry, perhaps most importantly the former Communist countries of eastern and central Europe, has led to significant differences among plants of similar design. These differences have been not only in plant concept and design but also in standards of regulation, and therefore in quality of maintenance and operation. In addition, the resources available to safety inspectorates in different countries have varied widely.

Moreover, there has been a tendency for each country, in pursuing its national nuclear policy, to insist on 'customizing' reactor designs, even when buying a design from abroad. This has had a number of results. First, economies of scale that might have been available from large, multinational production programmes have not been enjoyed, and investment costs have been higher than might have been necessary. Second, the cumulative operating experience of each reactor series has been limited, with potentially negative implications for operational reliability and safety.

As noted earlier, parallels with the aircraft industry are often drawn. In the aircraft industry, there are a small number of internationally licensed designs. There is no question of a design requiring different licensing for use in different countries, which would clearly be impractical for international air travel. Some commentators argue that the nuclear industry should evolve in this direction. Once a particular design had a licence in its country of origin, it would be deemed licensable in other countries.

There are difficulties with this approach. International markets in the higher-technology components already exist; but as these components are often the parts that add the most value to the plant, it is possible that governments, at least in the larger countries, will encourage local firms to build capacity. Licences are thus site-specific, and cannot easily be transferred from one site to another. In addition, sensitivities about nuclear technology may limit the freedom with which certain components are traded internationally, especially with respect to countries that are not signatories to the Nuclear Non-Proliferation Treaty. It is likely, then, that a considerable degree of national differentiation will persist. On the other hand, an international market in reactor components may well evolve as a result of consolidation in the nuclear construction market, with possible benefits for quality control and reliability – though it may take some time for such markets to develop for components of novel reactor designs.

Most commentators seem to agree that locally organized regulation is the most effective approach. The variety of reactor designs, organizational structures and management, training and operational frameworks that exists makes it unlikely that a single international regime can be applied effectively to them all. For instance, despite the similarities in nuclear technology, the design details of particular plants will vary, and this will depend on the local environment – local geography, such as the risk of earthquakes, and local political demands, the availability of different transportation links etc. The non-nuclear side of the plant is also likely to differ from company to company and from country to country.

Nonetheless, strengthening international links between operators and national regulators is likely to result in the sharing of best practice. The IAEA has published a series of safety standards that are widely applied, and the activities of WANO (referred to earlier) have been important in bringing together the operators of stations worldwide. Groupings of regulators with special shared interests, such as the VVER group meeting under the OECD, are increasingly common. In some cases, however, perhaps most notably India, the application of sanctions associated with the development of a military nuclear capacity acts as a barrier to international collaboration.

Care must be taken to allow regulators to retain their independence. There is a danger that should a government have a particular desire to see nuclear power expand or contract, this desire might become translated into pressure on regulators to be 'helpful'. The apparently intimate relationship between the industry and government in some countries, at least until recently, has undermined confidence among some pressure groups in the independence of regulators. It is also essential that regulators are given appropriate resources and legal powers to undertake the task.

Reactor designs

It is in the general nature of technologies that they tend to become safer over time, although of course the magnitude of external threats such as terrorism can rise and fall. Experience gained from incidents such as Three Mile Island have led to engineering improvements to evolutionary reactor designs, which, according to appraisals, have improved safety and operational performance. New large-scale reactors have features such as additional shutdown systems, a backup reactor-protection system, fourfold redundancy of primary safety equipment and more extensive protection against extreme environmental hazards, such as earthquakes. Much attention has also been paid to reducing the likelihood of any major accident scenarios remaining unforeseen, including those that could arise through common causes such as human error.

The debate about the merits of small, passively safe designs relative to larger designs, which require more engineered safety systems but benefit from economies of scale, has a long history. Recently, an interest in smaller reactors from an economic point of view has rekindled the discussion. It is argued that although smaller reactors do not enjoy the same economies of scale as larger reactors of similar design, the larger surface area-to-volume ratio, and other features associated with smaller reactors, reduces the requirement for engineered safety systems, with compensatory cost savings.

However, although reactor safety has almost certainly been improved in recent years, the possibility of large-scale off-site emissions cannot be eliminated entirely. Nor can calculated levels of safety be checked against operating experience, because enough operating experience can never be gained. Some commentators argue that increasing the complexity of the hierarchy of engineered safety systems may even introduce extra hazards. For example, the safety systems themselves might interact in unexpected ways, there could be unrevealed faults within them or the ability of operators to understand what is happening may be compromised by increasing complexity.

Large-scale reactors with a high level of engineered safety are likely to play a part in any future that includes new nuclear construction. In the Far East, several countries are continuing to develop their conventional pressurized water reactor and boiling water reactor technology base. In Europe, France has continued to develop enhanced PWR technology, which has shown less focus on passive safety.

In other countries, the future prospects of nuclear power may be aided by the adoption of designs with more passive safety features. Some recent generations of reactor concept have concentrated on enhancing intrinsic safety. Examples at various stages of conception or design include BNFL Westinghouse's AP600, GE's simplified boiling water reactor (SBWR), the next-generation CANDU and the Russian VVER-640. These designs seek to offer significant improvements in demonstrable safety through plant simplification and improved reliability. In the AP600, for example, the safety systems use natural forces such as gravity or natural circulation to enable the systems to cool the reactor adequately after an accident. Extensive efforts have been undertaken to ensure that the safety systems will perform as intended and deliver a high degree of passive safety. The designers BNFL Westinghouse claim that the net effect of this is that the theoretical predicted core-melt frequency is better by a factor of about 100 than that of a typical modern PWR.

A more revolutionary emerging development with major passive safety features is the Eskom 110-MW high-temperature Pebble Bed Modular Reactor. In this design the fuel comprises uranium in ceramic pebbles that are highly resistant to extreme temperatures. This very high-performance fuel is combined with reactor physics characteristics that inherently reduce the reactivity of the system as the core temperature increases. Although this design is at an early stage of development, it is calculated that the robustness of the fuel and the reactor physics would allow it to withstand all rapid increases in temperature that could arise from reactor faults such as failure to shut down the reactor or a major loss of coolant.

The likelihood of major reactor accidents may be considerably lower in these designs but, if there are many more reactors operating, the general result might still be an increase in incidents. Some commentators therefore suggest that a more radical approach may be necessary. This would have to ensure that a core could not melt or disintegrate under any circumstances and that any accident would be confined within the plant containment. This approach might involve much lower power outputs, a separate sub-critical assembly with an external neutron source, new approaches to the integrity of primary circuits and/or burial of the reactor. The demonstration of the feasibility of some of these concepts, often referred to as 'Generation IV', is still some way off.

The commercialization of nuclear operations

The trend towards liberalization of electricity supply markets has interesting implications for plant safety. It might be expected that operating in a competitive market might lead to an increased 'production imperative', as privatized firms face greater pressures to reduce costs in order to improve returns to shareholders. However, as noted earlier, there can also be many pressures to ensure continued production in non-commercial power systems, such as those that pertained in the Soviet Union.

The process of privatization and liberalization of a variety of formerly state-owned industries – electricity, gas, water, steel etc. – has been a particular feature of the British economy in recent years. There is evidence that the safety performance of privatized British companies has improved in the period after privatization at a higher rate than safety in companies that have remained in private or state hands. There is also little evidence that safety performance in privately owned nuclear power stations is any different from that in state-owned plants, for example in the United States.

There appear to be several business drivers for establishing and continuing to improve safety performance:

- companies generally have a strong incentive to ensure that no one is harmed by their operations;
- accidents, especially nuclear accidents, potentially cost a great deal of money – directly by harm to the individual involved, by damage to the plant, by disruption to production and by fines and compensation, and indirectly through investigation and prevention activities;

- the nuclear industry in particular is very much in the public eye and is highly regulated; any incidents are very newsworthy – a good safety record is essential to fostering or maintaining public acceptability;
- in the global market an organization's safety performance is increasingly viewed as an important factor when seeking contract work, for example in nuclear decommissioning.

The nuclear industry, and other industries, would thus argue that 'good safety is good business'.

Summary

The inherent difficulty of appraising and managing low-probability, high-consequence risks will remain a central issue in nuclear safety. There are considerable uncertainties involved in:

- verifying the calculated risk of a major release of radioactivity because of plant malfunction or external attack;
- estimating the pattern of contamination associated with the release in any particular case;
- determining the health effects of low-level radiation exposure.

The accident at Three Mile Island led to a closer relationship between US nuclear utilities in the field of safety. The accident at Chernobyl was similarly influential in improving the flow of information and expertise internationally, and in particular between the West and the former Communist countries of Europe. Chernobyl demonstrated that a major nuclear accident was a possibility, in at least some designs of plant.

The recognition that a major accident is possible is an important component of nuclear safety; its absence seems to have been an important psychological factor behind the Chernobyl accident. The strengthening of international co-operation and collaboration, through organizations such as the IAEA and WANO, has been a useful development that will bear fruit whether or not there is a major programme of new reactor construction. Should more reactors be built, there will be a focus on new approaches to safety, either through adding more engineering solutions or through simplifying and reducing the size of reactors and making use of more passive safety features. In reality, it is likely that both of these courses will be followed.

On balance, there is no compelling reason to believe that the increasing commercialization and liberalization of power supply markets will have a detrimental effect on safety, although regulators will have to be aware of pressures to cut costs. However, there is a feeling in many countries that more resources will need to be dedicated to nuclear inspectorates should a major increase in nuclear generation be envisaged.

References

Grimston, M. C. (1997), 'Chernobyl and Bhopal ten years on – comparisons and contrasts', in M. Becker and J. Lewins (eds), *Advances in Nuclear Science and Technology*, Vol. 24. New York: Plenum.

HSE (1992), *The Tolerability of Risk from Nuclear Power Stations*. Sudbury, United Kingdom: HSE.

IAEA (1996), *Defence in Depth in Nuclear Safety*. Vienna: INSAG-10, IAEA.

Ignatenko, E. I., V. Ya. Voznyak, A. P. Kovalenko and S. N. Troitskii (1989), *Chernobyl – Events and Lessons (Questions and Answers)*. Moscow Political Literature Publishing House.

Kemeny, J. G. et al. (1981), *Report of the President's Commission on the Accident at Three Mile Island*. New York: Pergamon Press.

United Nations Scientific Committee on the Effects of Atomic Radiation (UNSCEAR) (1988), *Sources, Effects and Risks of Ionising Radiation*. New York: United Nations.

6 Nuclear energy research, development and commercialization

Introduction

Experience in other energy industries such as oil and gas has shown that technologies require constant updating in order to stay competitive as new developments become possible and competing technologies become more cost-effective and improve in other ways. In a competitive area, improvements in one technology often spark similar improvements in others, and any technology that stands still will inevitably be overtaken. The nuclear industry is no exception. Indeed, its proponents point to significant improvements in the efficiency and safety of nuclear reactors over the past decade and claim that new approaches to nuclear generation, based on existing designs or involving more radical departures, now promise to deliver a much more effective technology. However, government aid may be necessary in bringing new designs to commercialization before organizations will have the confidence to order new plants.

Opponents of nuclear power point out that the 'significant improvements' have been necessary simply to increase the output and safety performance of individual stations up to their design levels, and should not mask the fact that many nuclear plants have performed very poorly over their lifetime. From this point of view, the vast sums spent on nuclear power research and development (R&D) over the past 50 years, provided mostly by taxpayers, have largely been wasted. Nuclear research should thus be restricted to developing techniques to clean up the legacy of waste and redundant plants, while the main R&D investment should be directed at renewable energy and energy efficiency, where it is regarded by many as likely to be more cost-effective.

The context of research expenditure

Research expenditure can be categorized into two basic areas. The first is work for improving the efficiency, cost structure, operating life and safety of existing operations. This is usually of a relatively short-term nature, and any benefits from it accrue to the owner or operator of a plant. There is some risk that the research will not prove cost-effective but, as often the work is in areas well known to the operator, the risk is rarely substantial. Crucially, success brings early benefits. Examples include the development of more heat-resistant fuel rods for nuclear reactors and improvement in burn-up rates, leading to lower fuel requirements and shorter refuelling outages. Some companies call this work 'technological development', to differentiate it from the other category.

The second category of research is work to develop new products or processes. This research is normally of a longer-term nature. It can involve initial laboratory investigation, followed by the construction and operation of one or several pilot plants of increasing scale and finally the construction of a commercially sized plant. Where complex plants such as nuclear reactors are involved, few power generation companies will take the risk of ordering a plant that has not been tried on a commercial scale. (When this was attempted in the 1960s and 1970s, it was found that scaling up from a relatively small-scale to commercial size could cause enormous problems taking many years to resolve.) Furthermore, experience has shown that the capital costs of a 'first' plant tend to be far higher than those of subsequent units.

When it comes to more modular types of technology, such as some of the renewables, the activity requiring time and funds may be not the scaling up but the demonstration of the necessary infrastructure – for example the power connections to offshore wind farms – under widely different political and economic conditions. This type of research can take decades to move from initial idea to commercial product, and there are no guarantees of success. Even at the end of a technically successful process, there is no guarantee that the process can be turned into enough profit to provide an adequate return on the capital and human effort involved. One common 'killer' is that better technology developed more quickly by rivals could nullify the assumed advantage of the basic idea.

This second category of research is far more risky than the first. Although the eventual returns can be enormous, no early profits can be expected. Nonetheless, during the period between 1945 and the mid-1980s many large corporations saw it in their interest to carry out long-term research, including some fundamental research of the kind more commonly carried out in universities. However, as competitive conditions became tougher in the 1980s and 1990s and shareholder pressures for high returns increased, this type of research became one of the first casualties in the race to cut costs. The fact that in a competitive market the required return on capital rose to well above 10 per cent increased the pressure to cut research budgets.

It could be asked why other high-technology industries are still able to spend their own funds on expensive and long-term research, the drug and aircraft industries being examples. Part of the answer may lie in the very high return on successful products, in contrast to the heavy regulation that faces electricity generators even in liberalized markets. It has been understood for a long time that the high profit margin of the successes, largely guaranteed through patent protection, makes the investment in research worthwhile and that any major reduction in margin may destroy this incentive. Although patent protection can be of importance in the energy industry, it will never be more than a minor factor: in most instances the developers of a new process have just a few potential customers, while the potential consumer base for a new drug can run into millions. In addition, the ultimate product, such as the kWh of power, is difficult to differentiate, unlike the potential application of a new drug for a particular condition.

The role of government in organizing and funding long-term research, development and commercialization (R, D & C) in a competitive market is an important but controversial one. When governments saw themselves responsible for ensuring secure supplies of energy – when, in a sense, energy was regarded as a social or industrial service – it was largely accepted that government also had the duty to ensure that the necessary technology was available to carry out this responsibility. Government laboratories were established in many countries in order to develop the necessary technology. But when responsibility for energy provision is seen to be vested in the market, it can be argued that government should no longer have this role, as the industry itself will be in a better position to determine what R&D is required. And because it will reap the benefits, industry should also provide the funding. However, as discussed, an industry working in a liberalized and strongly competitive market is unlikely to engage in long-term, speculative research.

There is therefore a dilemma over who should fund long-term energy research. In order to address this, it must be determined who will benefit from this research. A principal aim of energy R, D & C is to achieve the required restructuring of the energy sector without adversely affecting living standards to an unacceptable degree. It can be argued that the eventual beneficiaries of long-term research will be the 'public', who will enjoy more secure and more economic energy supplies and less environmental damage. The R, D & C then should be funded largely from general or specific taxation, at least to the point where key uncertainties have been reduced to the extent that industry will take up the research as a potential contributor to long-term profitability.

Research is often compared to insurance.[1] A company spends money on research in order to ensure that it does not become uncompetitive and lose its

---

[1] Schock et al. (1999).

markets. By spending funds on research, a government ensures its people against energy crises, global warming and air pollution. The costs of this work are analogous to insurance premiums. In fact, it can be argued that the case for R&D is even stronger than this. An insurance policy rarely compensates victims for the time, inconvenience etc. associated with their loss, while timely investment in R&D may head off problems before they arise.

However, expenditure on a particular R&D project does not bring guarantees of a payback in the way one would expect from true insurance. An alternative view, then, is that R&D creates options with uncertain benefits but that modern business analysis, including 'real options techniques', can be applied in estimating the benefits.[2] This view has its sceptics, who hold that as the results of applying these techniques can never be accurately evaluated, their usefulness may be limited.

Vine concludes that there are three categories for research:

- short-term research – the responsibility of industry;
- R&D requirements involving a vital national interest or projects that are very long term (for example, defence or nuclear fusion) – the responsibility of government;
- issues of common interest and benefit, such as competitive technology for energy – a joint responsibility of industry and government.[3]

Nuclear R&D must be considered in relation to expenditure on R&D for other energy sources, total government expenditure on R&D and developments in other fields. For example, a number of key developments in the energy field in recent years, most notably the combined cycle gas turbine (and perhaps, in the future, the small-scale fuel cell), have not originated in energy R&D.

A number of general questions arise. What is the purpose of technical research and development? Who should fund it and the subsequent commercialization? Can any lessons be learnt from the past? How much is going on, in the context of other energy sources? What are the longer-term options? How can international cooperation be encouraged? This chapter seeks to address these issues.

---

[2] EPRI (2001).
[3] Vine (2000).

Energy research, development and commercialization

Background

There are four main requirements from the energy industries in a developed economy:

- secure supplies;
- environmentally acceptable supplies;
- economic supplies;
- socially acceptable supplies (in particular, supplies that the public recognizes as 'safe').

From time to time, perceptions of threats to one or more of these requirements change. Concerns about the international security of energy supplies, for example, generally declined between 1980 and 2000, while fears of climate change increased (although as yet there has been little policy response). Responding to these changes in perception affects a number of aspects of government policy, including views of the most appropriate organization for energy markets. Broadly speaking, governments have accepted that they have a role to play in funding R&D, at least to counter threats to secure energy supplies and to respond to environmental and social problems associated with energy.

As perceptions of potential threats to the four energy requirements have changed in recent decades, so have the attitudes of many governments in developed countries towards sponsoring energy R&D. Between the early 1970s and the early 1980s the main driver was the fear of insufficient availability of oil and natural gas, partly owing to resource limitations and partly for political reasons. It was assumed that energy prices would remain high for the foreseeable future. Research was concentrated on promoting alternative energy sources to 'conventional' (Middle East and Texas) oil. These included deep-sea oil and gas production, the use of oil shale and heavy oil fractions, more advanced coal mining techniques and nuclear energy. There was also increased emphasis on energy efficiency, often supported by legislative measures. In general, government R&D expenditure rose substantially, doubling in the OECD countries between 1975 and 1980.

Between the mid-1980s and the late 1990s, fears about hydrocarbon fuel shortages and sustained high prices evaporated, as oil prices fell back to close to their pre-1970 levels and quoted oil and gas reserves increased significantly, for two reasons. There was a reappraisal of known resources, partly in the light of new technology, and there were some major new discoveries, especially of gas fields. When security ceased to be an issue of concern, attention moved towards the other requirements of the energy industries, especially economic efficiency. Many governments came to the conclusion that the introduction of competition into energy, notably electricity, markets was most likely to reduce costs of generation and distribution.

As liberalization of markets proceeded, so governments argued that energy R&D, as in other industries, was now the responsibility of commercial companies, which would make profits from exploiting new technologies. Governments' energy R&D spending plummeted in most developed countries, the exceptions (Japan and to a lesser extent France) being those most concerned about threats to the security of their energy supplies and in which (for the same reason) market liberalization progressed most slowly. The companies operating power stations now found themselves answerable to shareholders, who would often be unwilling to devote large resources to speculative research programmes. In view of the considerable over-capacity in power generation in many countries, a result of the oil-driven recession of the late 1970s, there was little incentive for companies to carry out longer-term R&D. Thus company R&D spending also fell.

During the past few years there has been evidence that the policy emphasis is changing again, with both governments and industry looking more favourably on R&D investment. It is now widely recognized that over the next decades the world will face severe challenges in the energy sector. In the first half of the century, energy demand may increase by a factor of two or three, largely owing to rapid economic growth in developing countries, and existing plants will have to be replaced. At the same time, greenhouse gas emissions may have to fall by as much as 60 or 70 per cent in order to combat climate change. Although either or both of these apparent problems may prove to be exaggerated, it would be highly risky to presume that they can be ignored – and they may prove to be underestimates. To address them both will require fundamental changes to the world energy market, which for 150 years has depended on ever-increasing use of fossil fuels. It is difficult to see how a technological 'business as usual' scenario could possibly achieve the required changes.

A reliable and affordable energy supply is a vital ingredient of a healthy economy. Experience in the 1970s and more recently in California shows that in the developed world, energy shortages, whether of electric power or motor fuel, are highly destabilizing economically and politically. Furthermore, nearly one-third of the world's population does not have any access to electricity, while many others in industrializing countries have only inadequate and unreliable supplies. Without far better access to energy in the developing world, there must be grave doubts whether the present wide gap in the standard of living between rich and poor countries can narrow. The persistence or widening of these inequalities carries the risk of international tension and perhaps of terrorism.

Meanwhile, the International Panel on Climate Change (IPCC) is issuing ever-stronger warnings about the risks of global warming unless emissions of greenhouse gases are radically reduced. It points to substantial rises in the average global temperature over the last century, a rise in the sea level and a significant reduction of snow and ice cover, claiming that there is 95 per cent likelihood that these changes are linked to human activity. Even in the absence of complete certainty, preventive action under the precautionary principle would appear to be vital.

The costs of meeting long-term energy and environmental targets

The international community has taken steps to respond to the need to control greenhouse gas emissions, although the process has inevitably been steeped in politics and has moved more slowly than many commentators would have preferred. One barrier to progress has been the presumption that, using traditional technologies, the costs of making major cuts in emissions could be sufficiently high to have a significant effect on standards of living in the industrialized world.

Energy R&D may make an important difference to this perception. Over the years it has delivered vast technological improvements to the energy industry, and there is no reason to believe that it could not continue to do so as long as adequate resources are made available to it. Examples from the past 10 to 15 years are plentiful. In oil and gas production, the installation of sub-sea wellheads, new imaging techniques (notably 3D and 4D seismic) and the use of horizontal and 'smart' wells have resulted in significant cost reductions and thus access to more oil and gas. Similarly, developments in gas turbine technology, largely because of defence needs, have revolutionized the use of natural gas for power production. In the field of bio-energy, sustained research funding, largely by the US Department of Energy, over a period of 10 years has reduced the estimated costs of bio-ethanol production from some $0.92 per litre to $0.32 per litre.[4] As possibilities for further reductions have already been identified, the process may soon be ready for the commercialization of ethanol and ethanol derivatives for blending into gasoline, thereby reducing the use of hydrocar-

---

[4] Wyman (1999).

bons, reducing the import of crude oil into the United States and reducing the net emission of carbon dioxide. (Ethanol is in effect a bio-fuel, the carbon dioxide released during its use being equivalent to the amount absorbed during the growing of the crops from which it is made.)

A more wide-ranging study of America's potential for reducing carbon dioxide emissions, by a working group drawn from five national laboratories, has taken place in the United States. Drawing on a detailed study on this subject in 1997,[5] the working group developed a report on 'Scenarios for a Clean Energy Future'.[6] Its main conclusions were:

- appropriate public policies could significantly reduce not only carbon dioxide emissions but also air pollution, petroleum dependence and inefficiencies in energy production and use;
- the general economic benefits of these policies appear to be comparable to their overall costs;
- uncertainties in the assessments are unlikely to alter the general conclusions.

The report predicted that the implementation of those policies could result in annual savings to the US energy bill of some $50 billion (some 8 or 9 per cent of that bill) by 2010 and $100 billion (or 15 per cent of it) by 2020 when compared to a 'business as usual' scenario. These savings, it was argued, should readily cover the costs of enhanced research and of bringing the research results to commercial application. In other words, over a period of decades there should be little or no cost to the US economy of following the path suggested by the working group, although in individual sectors there would undoubtedly be winners and losers. Furthermore, improved energy efficiency would improve US competitiveness, which in turn would force other countries to follow. (From 1975 to 1985, the energy intensity of the US economy – the amount of energy used per unit of economic output – fell by about one-third, although it rose again as energy prices fell.) Of course, were the United States to forgo these improvements, other industrialized countries might well take the lead and thus start the virtuous circle.

The ultimate aim of energy R, D & C should be to facilitate the required restructuring of the energy industries, say by mid-century, without adversely affecting living standards to an unacceptable degree. This could perhaps be

---

[5] USDOE (1997).
[6] Brown et al. (2000).

achieved best alongside changes to the regulatory and taxation regime, to encourage sustainable investment.

There are four clear options for reducing the emission of greenhouse gases:

1. improving energy efficiency (in transformation, distribution and end use);
2. using renewable energy;
3. expanding nuclear energy;
4. sequestration of carbon dioxide from the burning of fossil fuels.

Other new technologies, such as the fuel cell, may facilitate the implementation of one or more of these options, for example by allowing effective 'storage' of electrical output from intermittent sources. These options are often seen as being in competition. In practice, however, the scale of the task of restructuring the energy sector is so large that all four may well have to make a contribution. Because of the current uncertainties over the potential contribution that each could make, it would seem prudent to retain and develop all of them.

In 2000, fossil fuels made up some 90 per cent of the 8.75 Gtoe (billion tonnes oil equivalent) of global commercially sold primary energy. Should a target be set to reduce carbon dioxide emissions to 50 per cent of the 2000 level, say by 2050, and assuming that by then total energy demand will have doubled, the proportion of fossil fuel would have to fall to below 30 per cent of energy used globally (ignoring possible effects from sequestration, improved efficiency of use or switching from coal to gas). The resulting gap would have to be made up by the four options listed above. Of these, nuclear energy supplied about eight per cent of global energy in 1990 and 'new renewables' only a tiny fraction. ('Old renewables' encompass large-scale hydro and non-commercial fuel, such as dung and firewood. They met some 16 per cent of energy demand in 1990, and are expected to decline in future years as a proportion of total energy used globally.) Sequestration – the removal of carbon dioxide from flue gas and its disposal in such a way that the gas cannot escape into the biosphere – is a relatively new concept. It first emerged in the 1970s as a by-product of enhanced oil recovery using carbon dioxide, but was only used to a modest extent even in the 1990s. It is, however, the subject of a number of research projects, notably at the Statoil (Norway) operation in the Sleipner West gas field.

To achieve the necessary reduction in fossil fuel use by 2050 would require enormous contributions from the four options. This is not an impossible task, but an early start in developing the options would seem to be essential.

For the present, it is difficult to determine the level of contribution each option could realistically be expected to make over the next 20 to 50 years. A thorough appraisal is urgently needed. What return, in terms of effective new technology, could be expected from what level of R&D investment in each of the options? What special problems might face intermittent sources of energy, the integration of which into different infrastructures might be especially difficult? The nature of these questions, and the answers to them, may vary greatly between countries or even between regions of a country.

When it comes to 'heavy' technology, such as sequestration or nuclear energy, managing the finances of the demonstration phase becomes particularly demanding. It is unlikely that an individual company would wish to bear the costs of an expensive demonstration plant without very firm reasons to believe that the technology would be attractive in the medium to long term. Should there be considerable uncertainty in this area, demonstration may be possible only with government support, or possibly through a consortium of interested commercial parties. In any case there would need to be some indication that the technology would be acceptable to the public and to government. Past experience has shown that planning R, D & C in the area of energy requires close collaboration between government and industry in order to prevent the parties following contradictory paths.

Nuclear R&D in the context of total energy R&D

Table 6.1 is taken from the International Energy Agency's database regarding government energy R&D expenditure in the 26 IEA member countries. The database does not include information about private companies' expenditure or about funds spent by non-IEA countries such as China, Russia or India. The total amount of energy R&D expenditure by governments of IEA countries rose in tandem with the rise in oil prices in the mid to late 1970s, and fell away significantly as perceptions of shortages of hydrocarbon fuels declined subsequently. This trend is even more dramatic if the figures for Japan, which maintained its R&D expenditure through the 1980s and 1990s and accounted for over one-half of IEA expenditure in 1999, are excluded. The table shows a consistent drop in expenditure between 1980 and 1999. Energy R&D has also represented a falling proportion of total government R&D spending in recent years. For example, in 1997 the level of energy R&D in the United States was below 10 per cent of total (non-defence) government-funded R&D.[7]

---

[7] PCAST (1997).

Table 6.1 Expenditure by IEA countries on energy R&D
(1999 $ million)

	1975	1980	1985	1990	1995	1999
Conservation	351	977	745	530	1,081	1,141
Fossil fuels	602	2,580	1,528	1,769	941	524
Renewables	209	1,934	870	589	688	536
Nuclear fission	4,969	6,952	6,819	4,198	3,488	3,205
Nuclear fusion	611	1,226	1,487	1,080	1,003	686
Other	1,021	1,599	1,066	1,203	1,419	1,385
Total energy R&D	7,763	15,268	12,515	9,369	8,620	7,477
Total energy R&D (excluding Japan)	6,312	11,960	8,918	6,047	4,771	3,633

Source: IEA (2001).

Non-government R&D investment is more complicated. There are indications that the proportion of all R&D funded by business sectors may have risen, aided in part by growing tax breaks for companies carrying out R&D, although, the pattern varied considerably – in Japan the business sector provided 72 per cent of total R&D funding in 1999, against 67 per cent in the United States and 55 per cent in the European Union. However, in the energy field it appears that business funding for R&D has been falling alongside that of governments – for example by 40 per cent in the United States between 1985 and 1995.[8] More recently, there has been an apparent revival in non-government investment and interest in developing nuclear designs, for example by British Energy in the Canadian Next Generation CANDU reactor, and by Exelon and BNFL Westinghouse in the ESKOM Pebble Bed Modular Reactor.

As is often noted by opponents of nuclear power, expenditure on nuclear fission dominated the overall figures throughout the period, although it did fall from 64 per cent of the total in 1975 to 43 per cent in 1999. However, even these figures can be misleading. Table 6.2 shows that, with the major exceptions of Japan and France, government R&D expenditure on nuclear fission in the IEA countries fell significantly in the 1990s, to levels below that spent on renewables. By 1999 Japan alone was responsible for some 77 per cent of IEA R&D expenditure on nuclear fission; France accounted for more than half of the remainder. If the French and Japanese figures are excluded, fission R&D expenditure by the rest of the IEA countries totalled $318 million, against $403

---

[8] Ibid.

Table 6.2 Expenditure by IEA countries on fission R&D
(1999 $ million)

	United Kingdom	France	Japan	United States	Other IEA countries	All IEA countries
1990	188	385	2,407	644	574	4,198
1999	3	428*	2,459	19	296	3,205

* 1998 data.

Source: IEA (2001).

million for renewables. It can be argued, however, that as long as someone is doing longer-term nuclear R&D, there is no need for every country to do so.

Outside the IEA, Russia, India and China have substantial nuclear fission programmes, and as the European Union also funds an amount of fission R&D, the worldwide totals for fission may well be rather higher than the figure opposite. Nonetheless, given that the bulk of government-sponsored R&D into nuclear fission focuses on waste management and other fuel cycle back-end processes, it is clear that most governments have spent relatively little on new reactor designs in recent years. The emergence of projects such as the Generation IV nuclear energy initiative (described later) may presage a change.

For some of the renewables, the pattern of R, D & C funding differs from that of other energy sources. In the cases of nuclear fission, nuclear fusion, sequestration and perhaps renewables such as large geothermal or bioethanol facilities, large-scale prototype plants eventually have to be built. The cost of prototype facilities for renewables such as wind power, solar power or biomass is likely to be at least an order of magnitude less.

This in turn may increase the riskiness of R, D & C for nuclear power in comparison to many of the renewables. Diversifying the research effort over a large number of small projects is likely to lead to significant improvements in the technology in question. Reliance on a single large demonstration unit – all that is likely to be affordable given the cost – runs the risk of losing the whole investment if insuperable problems are encountered (although of course it also offers the potential for major advance). Experience of this kind in the past has contributed to a growing unwillingness, which has accompanied the liberalization of power markets in some countries, to sponsor 'big science'.

However, as indicated in a recent study,[9] renewables are likely to need temporary subsidies during the commercialization phase, until costs fall as large

[9] G-8 (2001)

numbers of units are manufactured and installed. It will also be necessary to gain experience of the infrastructure required by different renewables within different environments. The total sums required for renewables may therefore be substantial. Global investment in renewables may rise to between $200 and $600 billion in the period 2000 to 2010,[10] although of course only a small proportion of this will represent commercialization subsidies.

When past expenditures are looked at, there seems to be little evidence that there was an allocation for energy that was then split among different sources. Less expenditure on one energy source, therefore, does not necessarily imply more for another source, although this might become the case in a world with a more coherent approach to the issue.

The question remains, then, as to how much funding should be accorded to each of the four technical options for reducing greenhouse gas emissions. An answer would depend largely on estimates of the R&D requirements and the potential contribution that each could make. Although attempts have been made to address this issue for individual technologies, there does not seem to be an authoritative study that looks at the various options side by side. The application of modern business appraisal techniques to the range of energy R&D options might be fruitful. These approaches have their critics, but in the absence of any methodical appraisal, the danger remains that R&D will continue to be guided largely by short-term political considerations.

Main areas of nuclear R, D & C

The present low level of demand for new nuclear installations for power production has led to a reappraisal within research establishments of the barriers, real or imagined, to expansion. As discussed in the chapters on economics and on public perception, these barriers range from doubtful economics in many countries when compared to the use of natural gas to public distrust of nuclear power in general. Moreover, there has been overcapacity in electricity generation in most OECD countries. In due course this overcapacity will diminish unless new plants are built (or demand is reduced dramatically).

The possibility thus remains that at some point, nuclear energy might be an attractive option for helping the world to meet an increasing energy demand and to make large reductions in greenhouse gas emissions. However, it seems likely that this could happen only if nuclear power succeeds in addressing some of its own problems. Research is therefore needed to:

---

[10] DTI (2001).

- improve the relative economics of nuclear power;
- improve safety in a manner perceived and understood by the public;
- reduce waste production, especially of long-lived elements;
- improve fuel efficiency (which in turn reduces waste production);
- ensure resistance to weapons proliferation.

New types of reactor

Recently, the following nomenclature has been proposed for reactor designs, describing their four 'generations':[11]

- Generation I refers to the early prototype reactors, such as Shippingport, Magnox and Dresden;
- Generation II refers to the commercial reactors built up to the end of the 1990s, for example BWRs, PWRs, CANDU, VVERs etc.;
- Generation III (and III+) covers advanced reactors newly licensed (or ready for licensing in the near future), such as the AP600, ABWR, EPR, VVER-640 and the South African PBMR;
- Generation IV systems are to be ready for construction in the 2020s and in operation by 2030.

Evolutionary designs are generally far less risky than novel designs, but can still offer significant improvements over the previous generation. Generation III designs (with the exception of the more revolutionary PBMR) incorporate lessons learnt during the 1980s and 1990s from the construction, commissioning and operation mainly of light water reactors (but also of gas-cooled and heavy-water technologies) throughout the world. Many of the designs are ready for licensing (indeed, ABWR, AP600 and System-80+ already have regulatory approval in the United States), and promise lower capital cost, greater speed in construction, quicker commissioning and better economic performance. As these designs (with the exception of the ABWR in Japan) have not been utilized yet, the power industry may not consider them to be commercially proven. In other words, although the R&D associated with them has largely been completed, their commercialization has not. However, the nuclear industry is confident that the design principles, being evolutions of proven approaches, are well understood and that the power industry will soon realize this.

---

[11] USDOE (2001).

If financing can soon be found for the prototypes, Generation III reactors may well be commercially available during the second decade of this century. However, as the first unit of a kind is usually more problematic and more expensive than the following units, investment in commercial prototypes tends to be very risky. Companies are reluctant, especially in a highly competitive market, to go ahead with development unless very large gains are possible if the project is successful or unless the risks of building the prototype can be widely shared or underwritten.

It can be argued, then, that there is a legitimate role for governments, either separately or perhaps in international collaboration, to provide funding or tax support for the construction of demonstration plants. Some commentators believe that this would be appropriate, as much of the regulatory uncertainty that deters private sector investment in prototypes originates with governments.

Generation III designs often make use of two important principles – simplification, leading to lower construction costs, and modularization, leading to lower construction times – alongside other improvements. For example, the Next Generation CANDU reactor, which uses light water instead of heavy water for cooling and slightly enriched uranium instead of natural uranium as fuel, might, it is claimed, be able to use spent fuel from LWRs as feedstock. A more modular approach to construction could also reduce the construction phase to 48 months or less. Taken together, these changes (if achievable) could make a substantial difference to the relative economics of this type of reactor.

The South African PBMR, based largely on German and early British R&D, is a more novel design. It is a high-temperature, helium-gas-cooled reactor; its power is produced by a gas turbine operating within the helium stream. If successful, the technology could have a number of advantages, and may bring substantial orders from all over the world. For instance, it has a modular design, making units as small as 110 MW economically possible and reducing the costs and duration of construction. Its thermal efficiency should be significantly higher than that of LWRs (some 45 per cent compared to about 30 per cent), and there could be a greatly reduced volume of spent fuel per unit of power generated. There are some major technical issues to be addressed, such as a turbine working under severe conditions, which add to the economic risk of constructing a demonstration PBMR. However, the risks of the project are being shared between two South African concerns (the power company ESKOM and the Industrial Development Corporation) and BNFL in Britain, with interest from the major US power producer Exelon. It is notable, however, that the major partner in the scheme, ESKOM, is a large state-owned monopoly. China too has constructed an experimental high-temperature gas-cooled reactor.

The Generation IV nuclear energy initiative. There are a number of other advanced reactors under study or on the drawing board. Most of these schemes require substantial development, and may not be commercially proven before 2020–30. In addition to basic research, these projects may need a series of pilot and demonstration plants, to allow the study in depth of factors such as the effects of different types of fuel, waste and the means of disposing it and the potential effect on proliferation of the fuel cycle associated with the envisaged reactor or assembly.

To explore the opportunities for this research, the US Department of Energy has initiated international studies. Nine countries (Argentina, Brazil, Canada, France, Japan, South Korea, South Africa, the United Kingdom and the United States) are investigating possible future developments of nuclear energy. The main purpose of these international studies will be to delineate the next generation of nuclear energy systems, and for that purpose a 'Generation IV International Forum' (GIF) has been formed.[12]

The technological goals for the Generation IV study have now been agreed. They are:

- to provide sustainable energy generation that meets clean air objectives and promotes the long-term availability of systems as well as effective fuel utilization for worldwide energy production;
- to minimize and manage nuclear waste, and particularly to reduce the burden of the long-term stewardship of waste;
- to increase the assurance that the waste streams are very unattractive for diversion into weapons;
- to excel in safety and reliability;
- to ensure a very low likelihood and degree of reactor core damage;
- to eliminate the need for off-site emergency response;
- to have a clear life-cycle cost advantage over other energy sources;
- to ensure that financial risks are comparable to those associated with other energy projects.

By March 2003 the US Department of Energy proposes to provide Congress with a 'road map' that will evaluate potential nuclear energy concepts, select the most promising line for further development and define the required R&D for bringing the project to commercialization within the proposed time (up to 2020 or 2030). Initial work has concentrated largely on reactors, but the full

---

[12] Ibid.

fuel cycle would also have to be considered before the 'road map' can be finalized.

Among the concepts being considered are:

- the rival HTGR under development by General Atomic Corporation in conjunction with France, Japan and Russia;
- advanced light water reactor systems, such as IRIS. This is a modular design of 100- to 300-MW unit capacity; it has a longer core reloading schedule (five to eight years) and enhanced safety features and proliferation resistance;
- renewed interest in fast reactors, although not necessarily as breeders. (It is said that new designs of fast reactor are simpler and safer than thermal reactors and that they can destroy plutonium and other actinides better and more effectively than via MOx making use of LWRs. This may become important in view of the need to destroy surplus weapons-grade plutonium as well as material from civil reprocessing. Much of the work on FRs has made use of sodium cooling, but more recently another coolant, a lead–bismuth eutectic, originally developed for Russian submarine reactors, has caused interest. However, widespread use of plutonium fuels could lead to concern about proliferation;
- substantial advances in the design and use of powerful accelerators, making it possible to consider accelerator-driven systems (ADS). (In these systems, assemblies containing fissile material operate in sub-critical mode – they do not produce sufficient neutrons to keep the nuclear reaction going, and require additional neutrons from an outside source, such as an accelerator.) For the present, the interest in ADS is connected largely with the treatment of waste (see page 196), but it is possible that sub-critical reactors may have advantages over critical fast reactors, especially as regards fuel composition. On the other hand, they require an external source of neutrons, which may be expensive;
- a number of other, perhaps more specialized, reactors that are under development, including high-temperature reactors providing high-temperature heat for extraction of unconventional hydrocarbon reserves, the production of hydrogen from hydrocarbons and for use in water desalination (a process already demonstrated by the BN-350 fast reactor in Kazakhstan and the subject of an 11-member IAEA technical cooperation project). Both these uses are likely to become important in this century;
- the Russian very small reactors, possibly down to 15 MW(e), with sealed fuel lasting the lifetime of the reactor. They would be intended to provide the energy source for isolated areas.

The IAEA has also established the International Project on Innovative Nuclear Reactors and Fuel Cycles (INPRO). Its initial nine members – Argentina, Brazil, China, Germany, India, the Republic of Korea, the Russian Federation, Spain and Turkey – contribute either money or cost-free technical expertise. INPRO's focus is on identifying the needs and requirements of a spectrum of developing and developed countries and on contributing explicitly to the debate on the global acceptability of nuclear power.

Uranium and the fuel cycle

There is a need to review assumptions about the fuel cycle as well as to work on new reactor concepts. When in the 1970s it was believed that nuclear energy would expand rapidly, the availability of conventional thermal uranium reserves was regarded as a potential bottleneck. It was therefore assumed that a fuel cycle involving recycling uranium and plutonium from spent fuel and using the latter in a programme of fast breeder reactors (in effect using the otherwise useless uranium-238 in natural uranium) would be necessary. The overall effect would be to increase the amount of energy that could be extracted from a quantity of natural uranium by a factor of 60. In the event, exploration for uranium found many more deposits, the fast breeder turned out to be more difficult and expensive than had been assumed and nuclear energy failed to expand as rapidly as had been expected. As a result, there is now no shortage of uranium, nor is one foreseen; neither of the world's two operating fast reactors is operating in breeder mode.

This situation is unlikely to change unless there is a major expansion of nuclear power. But should a major expansion take place, questions over the availability of conventional uranium reserves will eventually re-emerge. And even in a more modest nuclear future, countries such as Japan and India, which have plans for a major role for nuclear energy but lack indigenous uranium reserves, may be concerned about long-term overdependence on imports.

In principle, there are three ways of extending fuel reserves for thermal nuclear stations without pursuing the fast reactor route:

- better use of uranium from existing reserves;
- development of new uranium reserves;
- use of alternative thermal reactor fuels.

One area of research interest associated with the first of these options is the possible use of lasers in separating isotopes. At present, two techniques are in use for the separation of fissile uranium-235 from natural uranium (which consists mainly of uranium-238) – gaseous diffusion and centrifuging. Although the latter requires far less energy than the former, in absolute terms it is still costly and energy-intensive. It is predicted that the use of laser technology would be far more energy-efficient, thereby increasing the amount of useful energy to be extracted from a given amount of natural uranium. This process would also reduce the cost of producing enriched uranium for reactor fuel, but there could well be substantial proliferation dangers if a simple and cheap way of separating isotopes were to be developed.

Undoubtedly, should the price of uranium increase, new reserves would be discovered, and the extraction of uranium from unconventional reserves such as uranium phosphates might become economically attractive. Moreover, better mining techniques continue to develop, allowing greater yields from existing reserves. Research has an important part to play in all of these areas, although it is likely that the funding would come from companies involved in the fuel cycle rather than from governments, as commercial returns might be achieved relatively quickly.

More speculative is the possibility of extracting uranium from sea water, which is being pursued in Japan. Should extraction of uranium from sea water become economically feasible, perhaps in association with desalination as world water supplies come under pressure, then the lifetime of thermal uranium reserves could increase by two orders of magnitude. Should a major revival in nuclear power be in prospect, uranium from seawater could well be an important area of R&D.

India is looking at the possibility of utilizing thorium, of which it has large reserves, as the main fuel for its nuclear programme. As mentioned in Chapter 4, on waste management, thorium (unlike uranium) has no fissile isotope, but its main isotope, Th-232, is fertile because on absorbing a neutron it decays into the fissile U-233. Extensive research would be needed before a realistic comparison could be made between the advantages and disadvantages of the thorium fuel cycle and those of the uranium cycle. Most commentators believe that uranium will remain the basis of civil nuclear energy in most countries.

Alternative fuel cycles and reprocessing

As noted earlier, between the mid-1950s and the mid-1970s it was assumed that reprocessing plants would be required to recover plutonium from spent reactor fuel and that fast breeder reactors would be used to extract the maximum possible energy from natural uranium reserves. However, the United

States and some other countries abandoned this fuel cycle, because of fears that its use might greatly increase the risk of proliferation.[13] Since then, they have used a 'once-through system': the spent fuel that arises is classified as high-level waste, which, after some time in on-site storage while its rate of heat production falls, is disposed of in deep waste repositories. (In 2001, the National Energy Policy Development group recommended a re-examination of reprocessing options in the United States.) Perhaps an associated reason for the move against reprocessing in some countries was a recognition of the difficulties associated with the development of the fast breeder reactor and of the complexity and high costs of reprocessing plants.

A number of other countries – Japan, France, Russia, China and India – did not follow the United States' lead; they are still officially committed to developing reprocessing plants and fast breeder reactors. But at present only the French prototype Phénix and the Russian BN-600 fast reactors are in operation, and neither is being used to breed fresh plutonium. The world's only commercial-scale fast reactor, the 1200 MW French Superphénix, was closed in 1998. The Monju prototype plant in Japan, although closed at present, is expected to reopen for experimentation, and the 400 MW Fast Flux Test Facility at Hanford in the United States has been kept ready for possible recommissioning.

Current reprocessing plants use modernized versions of the Purex process, originally developed for nuclear weapons production as part of the Manhattan Project in the 1940s. As discussed in Chapter 4, today's rationale for reprocessing is quite different – and highly contentious – and there are doubts as to whether the present process, which is complex, expensive and generates aqueous waste streams that are difficult to handle, would be appropriate should reprocessing or partitioning (discussed later) be required in the future. At present only Japan is constructing a reprocessing plant, and it seems unlikely that the commercial plants in the United Kingdom and France will be replaced when they reach the end of their working lives. A revival of the case for reprocessing, presumably because of a significant growth in nuclear generation and a failure to develop new thermal fuel resources, might require the use of new approaches to reprocessing. Initial research has been carried out into novel approaches to separation, ranging from new aqueous processes to pyrochemical and electrochemical methods, mainly in the context of partition and transmutation. More R&D would be required to determine their suitability for large-scale reprocessing.

---

[13] Keeny (1977).

Managing radioactive waste

Managing radioactive waste has been called the Achilles heel of the nuclear industry. Early on it was assumed that high-level waste would be disposed of in deep underground repositories that would have to be kept safe for 100,000 years or longer. The first task was to find a site that was technically suitable and acceptable to the public. Thirty years later, no such site has been developed anywhere, although progress has been better in some countries, notably Finland and the United States, than others. During this period the costs of investigation have soared – in 1982, technical site characterization was projected to cost some $80 million per site in the United States; on the basis of experience at the Yucca Mountain site, the estimated cost has now increased to $5 billion.[14] In the United Kingdom a proposal in 1997 for an underground research facility as part of a site characterization was rejected because of doubts about the scientific methods used in the analysis. There is some debate as to whether the uncertainties typically involved in long-term waste management are amenable even in principle to scientific research.

While this argument rages, spent fuel and high-level waste from reprocessing is being stored mostly in surface storage on the sites where it was produced, and available storage is getting short in a number of cases. With the possible exceptions of Olkiluoto in Finland and Yucca Mountain in the United States, it is unlikely that underground repositories will be available within 20, perhaps even 30, years. (However, the events of 11 September 2001 may change the perspective on this issue.) The only possible solutions are to build more intermediate storage – on- or off-site, on- or near-surface – or to close down reactors as their storage becomes full. This latter course would imply the abandonment of nuclear energy, and is unlikely to be acceptable to a number of countries.

The question then arises as to what the potential lifetime of these stores should be. Bearing in mind today's uncertainties about the availability of final repositories and, indeed, the future scale of nuclear energy in the world, it may make sense to plan for intermediate storage with a relatively long lifetime. This could be for, say, 100 years, in the expectation that within that period the future of nuclear energy will have clarified and that there will have been more technological advances towards resolving the long-term waste problem. Research must make an important contribution to this debate because information will be required about the behaviour of the waste over time and

---

[14] Loux (1998).

about what conditioning, if any, and packaging the waste should receive. This behaviour is likely to vary according to the origin and age of the waste.

Should nuclear power be in global decline, an important question would arise about how to retain the necessary levels of expertise for dealing with waste within the industry for the several decades – or longer – between the closure of the last power plants and the eventual disposal of waste materials, some of which will arise only during the time of eventual site clearance.

Partition and transmutation

The purpose of partition and transmutation (P&T) is to transmute components of waste streams with long half-lives (mainly minor actinides) into elements having far shorter half-lives and thus to reduce the volume to be disposed of in deep repositories. This should make final disposal simpler. Venneri et al. suggest that it should be possible to reduce the time necessary to keep repositories secure to less than 500 years and to reduce the quantity of waste going to the repository by a factor of 25.[15] The process consists of separating the waste into a series of different streams, one of which contains all the material to be transmuted in an accelerator-driven assembly or fast reactor. The resulting spent fuel would again be partitioned into material to be recycled and material to be sent to a repository after conditioning.

Information from the United States[16] indicates that research into the various processes for P&T would take some six to eight years and cost about $280 million but that the total R, D & C might cost approximately $11 billion, with the first prototype being ready within 20 years. This may lead to a logjam, similar to those found elsewhere in the nuclear debate. P&T may make sense only in the context of an expanding nuclear industry, but the nuclear industry may be able to expand only if a waste management approach such as that offered by P&T becomes available. Collaboration among those working in this field, including the United States, France, Japan, the EU and Russia, may be a way forward.

International cooperation

As mentioned earlier, there is a strong case for the view that a dedicated application of R, D & C might well be one of the best and least costly ways of

---

[15] Venneri et al. (1998).

[16] USDOE (1999).

meeting the global demand for more energy while ensuring substantial reductions in greenhouse gas emissions. However, the effort involved in achieving this, which will involve work in all the four options discussed in Chapter 1, will be very great. There must be doubt whether the present system of funding energy R, D & C largely by individual governments and private companies will be able to deliver. Furthermore, it would seem wasteful to have teams in different countries working on largely the same long-term research ideas when a greater degree of coordination could reduce the duplication of effort.

In recent years support has grown in a number of quarters for more organized international cooperation,[17] and several such initiatives are now under way. In addition to the Generation IV group discussed earlier, the International Energy Agency, the Nuclear Energy Agency and the International Atomic Energy Agency have combined in a three-agency study called *Innovative nuclear reactor development: opportunities for international co-operation*. It emphasizes collaboration on 'enabling technologies' that may be of value in a range of possible future reactor designs. The International Energy Agency also has a wider remit – to foster international collaboration on more general energy R&D, for example through the International Energy Agency's Greenhouse Gas R&D Programme. The aims of this wider programme[18] include:

- dissemination of information;
- development of new products;
- development of networks;
- reducing costs of technology development;
- influencing the market;
- policy development.

It has been suggested that an international fund under professional and experienced management be created, perhaps connected to the World Bank. It would be of sufficient size to stimulate the development of technology for transforming the world's energy economy by mid-century, but without adversely affecting living standards to an unacceptable degree. This approach could run in parallel with existing international efforts arising from the Rio Conference of 1992, and could provide the necessary technological underpinning for those efforts. The strategy would aim at an accelerated development and deployment of the four options available for transforming the energy economy

---

[17] For example, PCAST (1999); Royal Society–Royal Academy of Engineering (1999).
[18] Koch et al. (2001).

and reducing dependence on conventional carbon-based fuels in all the main energy-using sectors (heat, power and transport). Wherever possible, the strategy would integrate government sponsorship with private sector initiatives and finance.

These initiatives would undoubtedly be difficult to implement. As technologies came closer to commercial exploitation, each participant would want to make sure that it gained competitive advantage over its erstwhile partners. Interestingly, some sceptics and some supporters seem to agree that international collaboration may be most effective on longer-term, speculative issues, while concepts closer to commercial viability would best be carried forward at national or company level. (Sceptics also point to a danger that 'funds will be wasted on projects to which everyone can agree because they do not represent commercial threats'.) Perhaps the Generation IV initiative will allow an evaluation of these issues.

Summary

No longer is the developed world, except notably Japan and France, dedicating very large sums of public money to the research of new designs of nuclear power station, as it did in the 1960s and 1970s. Expenditure on nuclear R&D has fallen considerably, and the focus is now on managing the legacy of waste from those early days and from current commercial nuclear operations. Many governments now regard nuclear power as a 'mature' technology, capable of providing resources to ensure its own long-term development.

However, most nuclear stations in operation, or even under construction, today depend on a technology that grew up in the days of centralized power markets. The high costs of installation could be amortized over long periods of time, there being a captive market for the electrical output. This cost profile sits uneasily in the competitive markets that now prevail in many developed countries. For nuclear power to compete in a competitive environment, it seems almost certain that reactor designs with lower initial costs and shorter construction phases will be needed. Other issues, such as achieving levels of safety that can be demonstrated to public satisfaction and developing acceptable waste management strategies, will also be important. And all of these developments will require research, development and, especially, demonstration.

With the return of concerns about global energy security and the growth of fears about climate change, nuclear power has again become an issue of public debate in a number of developed countries. Several international projects have been initiated to look at longer-term, more 'revolutionary' technologies, and the evolution of existing reactors has led to several new designs that are at or near the point of licensing. These include the advanced boiling water reactor, the AP600 and the AP1000, System 80+, the Next Generation CANDU and the Russian VVER-640 (alongside more speculative concepts such as the South African Pebble Bed Modular Reactor). These designs tend to focus on simplification (to reduce capital costs) and modularization (to reduce construction times) and on a variety of other improved features such as higher fuel burn-up.

Even if, on paper and in the laboratory, a new reactor design promises lower construction costs and more reliable operation, a potential investor will require hard evidence that these promises will be met in practice. This leads to something of an impasse. There may be little demand for new nuclear stations, at least in some areas of the developed world, unless the above problems can be overcome. But the research, development and commercialization investment needed to overcome the problems will not be forthcoming from the private sector unless there is a reasonable prospect of market demand for the new types of reactor.

The issue then arises as to who, if anyone, should be responsible for providing the resources for constructing demonstration plants. This may be of major importance in the future development of nuclear technology. Should this role be left solely to the companies that stand to make a commercial profit from the exploitation of new designs or is there a role for state support – either in direct grants or through other mechanisms such as tax breaks – to aid the demonstration phase? It should be noted that this dilemma may also apply to the development of large-scale renewables and to methods of sequestering carbon dioxide.

In competitive markets, it may remain difficult for private companies to commit the large sums necessary to demonstrate new nuclear reactor technologies or other techniques such as the partition and transmutation of waste. For the nuclear option to remain open, then, it may prove necessary for governments to offer some degree of support for building demonstration plants of evolutionary or revolutionary design.

References

Brown, M. A., M. D. Levine and W. Short (2000), *Scenarios for a Clean Energy Future*, Office of Energy Efficiency and Renewable Energy. Washington, DC: US DOE, *www.ornl.gov/ORNL/Energy_Eff/CEF.htm*.

DTI (UK Department of Trade and Industry) (2001), *New and Renewable Energy – Prospects for the 21st Century*, *www.dti.gov.uk/renew/condoc/energy.pdf*.

EPRI (2001), *The Value of Coal R&D to the United States, www.coal.org/epri.htm.*

G-8 (2001), *Report of the G8 Renewable Energy Task Force*, July 2001; *http// www.renewabletaskforce.org.*

IEA (2001), *Energy Technology R&D Statistics*. Paris: IEA, *www.iea.org/stats/files/rd.htm.*

Keeny, S. M. (1977), *Nuclear Power Issues and Choices: Report of the Nuclear Energy Policy Study Group*. Cambridge, Mass.: Ballinger.

Koch, H.-J., H.-J. Neef, I. Walker and K. Nakamura (2001), *The IEA's Collaborative Programme on Energy Technology – Developing a Bridge to a Low-Carbon Future*. Paris: IEA.

Loux, R. (1998), *Yucca Mountain – Follow the Money*. Carson, Nevada: Nevada Governor's Office Agency for Nuclear Projects, *www/state.nv.us/nucwaste/yucca/loux05.htm.*

PCAST (1997), *Federal Energy Research and Development for the Challenges of the 21st Century*. Washington, DC: PCAST, *www.ostp.gov/Energy/index.html.*

PCAST (1999), *Powerful Partnerships: The Federal Role in International Cooperation on Energy Innovation*. Washington, DC: PCAST, *www.ostp.gov/html/P2E.pdf.*

Royal Society–Royal Academy of Engineering (1999), *Nuclear Energy: The Future Climate*, ISBN 0 85403 526 5. London: Royal Society.

Schock, R. N., W. Fulkerson, M. L. Brown, R. L. San Martin, D. L. Greene and J. Edmonds (1999), 'How much is energy research and development worth as insurance?', *Annual Review of Energy and Environment*, 24, pp. 487–512.

USDOE (1997), *Technical Opportunities to Reduce US Greenhouse Gas Emissions*. Washington, DC: USDOE, *www.ornl.gov/climate_change.*

USDOE (1999), *A Roadmap for Developing Accelerator Transmutation of Waste Technology*, DOE/RW-519. Washington, DC: USDOE.

USDOE (2001) *Generation iv Nuclear Energy System, http:// gen-iv.ne.doe.gov.*

Venneri, F., N. Li, M. Williamson, M. Hours and G. Lawrence (1998), *Disposition of Nuclear Waste Using Subcritical Accelerator-Driven Systems: Technology Choices and Implementation Scenario*, LA-UR 98-985. Los Alamos National Laboratory.

Vine, G. (2000), *Federal Government Collaboration with the Private Sector for Energy Research and Development*. Palo Alto, California: EPRI.

Wyman, C. E. (1999), 'Biomass ethanol: technical progress, opportunities and commercial challenges', *Annual Review of Energy and Environment*, 24, pp. 189–226.

7 Nuclear power and the Kyoto Protocol

Introduction

The fact that nuclear power does not contribute significantly to emissions of gases associated with climate change, notably carbon dioxide and, less importantly, methane and nitrous oxide, has been advanced by its supporters as a major argument for further nuclear investment. Advocates argue that from this perspective, nuclear power is a strong candidate for a sustainable future. Opponents tend to argue that there are better ways of reducing greenhouse gas emissions and that nuclear power's other environmental consequences should be taken into account.

The physical phenomenon known as the greenhouse effect, whereby certain gases in the earth's atmosphere trap solar energy, resulting in global temperatures more than 30 degrees warmer than they would otherwise be, has been recognized for over a century. However, serious international concern about the climatic effects of releasing vast quantities of greenhouse gases into the atmosphere is a much more recent development. A European Community directive in 1990 sought to ensure that emissions of carbon dioxide, the main greenhouse gas, from the EC as a whole would be no higher in 2000 than in 1990. The UN Convention on Environment and Development (UNCED), held in Rio de Janeiro in 1992, and the subsequent conferences of parties (CoP) and meetings of parties (MoP) are a global attempt to address the problem. The 'Rio Convention' (the UN Framework Convention on Climate Change) is a key document in this effort.

Rio and its successors are a modest start to mitigating climate change. In broad terms, some 15 per cent of today's annual global energy use of 100 gtoe is provided from non-carbon-emitting sources. To stabilize the atmospheric concentration of greenhouse gases at double the pre-industrial level by 2050, some 50 to 70 per cent of annual global energy use of between 200 and 300 gtoe will have to be produced from non-carbon-emitting sources. These might include nuclear power, renewables – particularly biomass, wind and solar – or fossil fuel use with efficient capture and storage of the resulting carbon dioxide. Estimates suggest that reductions in carbon dioxide emissions of the order of 60 to 80 per cent of 1990 levels will be necessary in the developed world in order to achieve this stabilization. Even if it were to be achieved, a doubling of the pre-industrial level of greenhouse gas concentra-

tion would, on present projections, be associated with a significant distur-
bance of the global climate.

The stated objective at Rio that developed countries should emit no more
carbon dioxide in 2000 than in 1990 has been breached by most countries. And
even the provisions of the Kyoto Protocol of 1997 envisage only a reduction
of 5.2 per cent from the 1990 levels of emission of the six greenhouse gases by
the developed countries by the 'first' compliance period, in 2008–12. (This
expectation was much diluted at the Conference of Parties at Bonn in 2001.) A
regime of penalties to be imposed on those parties to the Kyoto Protocol that
fail to fulfil their obligations is now emerging, but its details will depend on
the establishment of subsequent compliance periods. In any case, even should
the commitments in the Kyoto Protocol be met, global greenhouse gas emissions may be some 30 per cent higher in the first compliance period than in
1990, as developing countries, whose emissions are growing significantly,
have no targets under Kyoto.

Alternative approaches

The process that began at Rio and culminated in the Kyoto Protocol has been
controversial, both in the details and in the principles. The failure of the United
States, responsible for about one-quarter of global carbon dioxide emissions,
to agree to the Kyoto process has been both bewailed and criticized in many
circles.

The US position has some merit, however. It raises three significant concerns about the Kyoto Protocol:

- the absence of emission targets for developing countries, expected to become the major source of greenhouse gas emissions over the next two or
  three decades;
- the cap on 'trading' emission reductions, which was promoted especially
  by the European Union, whereby a major part of necessary emission reduction has to be achieved by domestic measures, even if investing in
  projects in other countries might result in greater emission reductions;
- the focus on short-term emission-reduction measures.

In the view of the United States, the European position has moved from seeking to reduce concentrations of greenhouse gases in the atmosphere towards
trying to dilute that commitment with other goals. The European Union's
negative attitude towards nuclear power is cited as an example.

The view of the European Union, supported by many commentators, including numerous NGOs, is that climate change cannot be addressed properly without at the same time addressing other important issues such as the equitable distribution of responsibilities. Developing countries, it is argued, cannot be expected to limit their economic activity and greenhouse gas emissions unless the developed world, with much higher emissions, takes action to narrow the gap.

It does seem clear that the placing and timing of investment in reducing greenhouse gas emissions makes a considerable difference in the cost-effectiveness of the measures involved. As discussed later, directing emission reduction efforts to areas where existing plant or processes are most inefficient is more effective than directing the same effort towards countries that are already highly energy-efficient. Furthermore, to replace capital plant at the end of its operating life with more efficient versions or to divert resources to areas where there is rapid growth in demand for additional plant is more cost-effective than retiring operational plant, with some years of potential operation left, on purely environmental grounds. This effect would be strengthened if the time from now to the natural retirement point of the capital plant were spent in developing more efficient technologies than those currently available.

It would seem to follow that R, D & C programmes of appropriate size for the scale of the energy challenges ahead, coupled with a determined effort to replace plant coming to the end of its operating life with the best available technology, may well produce a better long-term outcome in terms of emission reductions than one that focuses on short-term targets. But whether countries would find it easy to pursue this course without the discipline of Kyoto-style compliance periods and concomitant penalties for failing to comply is a genuine question.

Whatever the merits of these arguments, however, it does appear that the Kyoto Protocol, as amended at subsequent conferences of parties to the Rio Convention, will be the main focus of global efforts to reduce greenhouse gas emissions for the foreseeable future.

Market instruments

As considered in Chapter 3, on the economics of nuclear power, the growth in concern about climate change has coincided with a period of liberalization of energy markets, notably in electricity, in many developed countries. From the beginning of the climate change negotiations, then, there has been discussion about introducing market instruments to aid emission limitation and reduction. This would be in preference to a simple 'command-and-control' model such as that which, a decade earlier, characterized European (but not American) measures to reduce the emission of sulphur dioxide and other gases associated with acid rain.

However, the climate change issue is not readily amenable to a 'polluter pays' approach. In theory, market measures to reduce pollution are effective where the cost of the instrument is equivalent to the cost of reducing the damage, or:

$$\text{marginal benefit} = \text{marginal reduction cost}$$

In the case of climate change it is impossible to determine the marginal benefit of reducing emissions owing to both the highly uncertain nature of the damage likely to be caused and the long time lag between the emission of the pollutant and the manifestation of the effects. As a result, it has been necessary to set some limits to emissions– inevitably arbitrary, however well they may be based on current scientific understanding – and to attempt to reach them at minimum cost. Initial discussions of how to achieve this gave consideration to global, regional or national carbon taxes, and again the European Union was in the forefront of the discussions. It was assumed that each developed country would be primarily responsible for limiting or reducing its emissions, through domestic action.

Flexibility mechanisms

In the immediate aftermath of the Rio Convention, attention began to be paid to mechanisms that would allow more flexibility in reducing greenhouse gas emissions rather than simply give each country a target to meet through reducing its domestic emissions. Consider the simplest case, of two neighbouring countries with similar emissions of carbon dioxide. In country A the transportation industry is extremely fuel-inefficient, while in country B the transportation industry is more efficient. It would make more sense for both countries to invest in improving the efficiency of transportation in country A than for each to make similar reductions domestically, as it would be much more cost-effective to put resources into improving the less efficient system.

The idea of internationally traded emission permits, allowing countries that reduced their emissions beyond their national targets to sell the surplus savings to others, was suggested. In addition, consideration was given to the concept of 'joint implementation' (JI). A government or commercial entity could invest in an emission limitation or reduction scheme in another country and gain benefit from certified emission reduction credits (CERs). These CERs could then be used to offset carbon tax or traded emission liabilities within the investing country. At first many developing countries (for example Algeria, Colombia and Malaysia), as well as some developed countries and environmental NGOs, opposed attempts by developed countries to extend JI to developing countries. Others, notably Costa Rica, were more positive. However, at CoP-1 in Berlin in 1995 a pilot scheme, known as Activities Implemented Jointly (AIJ), was agreed: there would be no restriction on which countries could act as hosts but no award of emission reduction credits in the initial phase.

Participation in AIJ has been rather disappointing, in part because of the lack of incentives offered by the scheme. As of February 2002, 156 projects were registered under AIJ. Of these, 77 (which would now be classified under the Clean Development Mechanism (CDM) discussed later) involved developing countries (see Table 7.1). Of the 156 schemes, 117 were sponsored by agencies within just three countries – Sweden, the United States and the Netherlands. In the field of energy, AIJ options include new power or combustion plant technologies, energy efficiency, and capture and storage of fugitive emissions.

In the time leading up to the Kyoto Conference of Parties (CoP-3) in 1997, it was still assumed that the mature version of AIJ would be largely, if not exclusively, carried out involving two or more developed countries with emission limitation or reduction targets. However, a number of countries argued for investment to play a role in developing countries in reducing projected

Table 7.1 Activities Implemented Jointly, pilot phase

Type of scheme	Total number of AIJ projects	(CDM)Projects involving developing countries
Afforestation	5	4
Agriculture	2	2
Energy efficiency	62	20
Forest preservation	9	8
Reforestation	5	4
Fuel switching	11	8
Fugitive gas capture	9	6
Renewables	53	25
TOTAL	156	77

Source: UNFCCC (2002).

emissions. Brazil proposed that developed nations should be fined for failing to fulfil their emission obligations and that the money raised should be placed in a clean development fund, to be spent in the developing world. Under pressure from the United States and Costa Rica in particular, this idea evolved into the Clean Development Mechanism (CDM). Money raised through the CDM would still be spent on emission limitation or reduction in the developing countries, but funding would come through the issuing of emissions credits rather than through fines. The CDM is often known as the 'Kyoto surprise'.

The Kyoto mechanisms

The conflict between those countries and organizations that want maximum flexibility (geographical, technical and temporal) in achieving emission limitation and reduction and those that believe that the bulk of emission controls should be achieved domestically by the biggest polluters remains the fundamental dispute at the climate change negotiations.

At the one extreme, and championed by the United States before it withdrew from the Kyoto process, is the argument for full flexibility. If a country can bridge the gap between its actual emissions and its emission targets by buying emission permits from abroad or by investing in emission reduction projects overseas (including protecting forests that would otherwise be felled or planting new forests), then it should face no restrictions in doing so. The opposing position, taken by some developing countries, the European Union and NGOs, has been that developed countries should reach their targets principally by implementing measures to reduce emissions domestically, with international trading mechanisms playing only a complementary role.

The debate at Kyoto and beyond has to a considerable extent been dominated by attempts to find a compromise between these two positions. The flexibility mechanisms that form part of the Kyoto Protocol can be viewed in this light. These mechanisms are:

- bubbles;
- tradable emission permits;
- joint implementation;
- the Clean Development Mechanism.

The first three will apply only to those countries with emission limitation or reduction commitments, that is the developed countries in Annex I of the Kyoto Protocol.

'Bubbles' will allow groupings of developed nations, such as the European Union, to pool their emission reduction targets and distribute necessary measures internally. Thus within the EU, Germany and the United Kingdom have significant reduction targets, while less economically developed members such as Greece and Portugal will be allowed emission increases. Tradable emission permits will allow developed countries with high compliance costs to buy permits from other developed countries with lower costs, the price of the permit being somewhere between the compliance costs in the two countries involved. It is likely that permits will be allocated on the basis of emissions in 1990. Joint implementation refers to projects funded completely or partially by one developed country in the territory of another, with credits for reducing emissions shared between participants.

The Clean Development Mechanism, although it resembles JI in some respects, is fundamentally different from the other Kyoto mechanisms. As noted earlier, the CDM would involve companies from Annex 1 (developed) countries investing in emission reduction schemes in developing countries. The host countries (by definition) do not have any emission limitation or reduction targets. Therefore, any emission reduction credits that are generated in the course of a CDM project will be used to offset emission control targets in developed countries, and will be additional to the credits or permits in circulation within those countries. In most cases the host country will not benefit directly from CERs generated by a CDM project, having no emission reduction or limitation targets of its own. (It is feasible, however, that certified emission reduction credits might in some way be shared with the host country, allowing the host to raise funds by selling them to agencies in developed countries). Other incentives must therefore be made available to the host country, notably financial aid for development. This is noted in Article 12 of the Kyoto Protocol, which introduces the CDM.

Principles underpinning the CDM

The Kyoto Protocol and its associated process have been deeply political in nature. Climate change was, at least in theory, the principal subject of concern. However, as noted earlier, the sectional and partisan interests of various stakeholders were also important. As a result, it is sometimes difficult to determine what principles underlie the Clean Development Mechanism.

However, it is clear that there are other, general issues, not specifically nuclear in character, that are relevant to deciding whether nuclear power should have a role to play in the CDM. They include:

- *National sovereignty.* There is agreement that no CDM project should be imposed on a host country without its explicit agreement. However, there is more dispute as to the extent that potential host countries should be denied certain technical options, even if those options would accord with the host country's own priorities for development and its interpretation of sustainability. Most obviously, the emergence of the concept of positive or negative lists (that is technologies that would be explicitly allowed under the CDM or those that would be explicitly forbidden) has led some countries to claim that this is an infringement of their right to determine their own technological future. In contrast, other states and some NGOs argue that investor countries, and perhaps the CDM institutions themselves (representing the wider international community), also have a right to decide what is 'sustainable', especially if the technology in question has international implications, for example the potential for weapons proliferation or accidents having cross-border effects. They have pointed out that the exclusion of nuclear power from the CDM does not prevent developing countries from taking it up but simply withholds an effective subsidy.
- *The role of the CDM executive board.* Paragraph 4 of Article 12 of the Kyoto Protocol states:

The clean development mechanism shall be subject to the authority and guidance of the Conference of the Parties serving as the meeting of the Parties to this Protocol and be supervised by an executive board of the clean development mechanism.

However, the role of the executive board has not been closely defined. It is conceivable, for example, that a particular type of project could be eligible for CDM accreditation in one context but not in another. If, as a case in point, a country has a well-developed programme of installing small-scale hydropower, a new scheme of that kind may be deemed ineligible for CERs, because it would probably have happened anyway. This might not be the case in a country where the technology was virtually absent. Similarly, technological progress might mean that a scheme that appeared unsustainable today might not appear so in the future.

These considerations suggest that the executive board should have much flexibility in coming to decisions about the appropriate treatment of individual proposals. However, this appears to cut against the concept of either positive or negative technology lists, and may also raise difficult issues of 'fairness' if one developing country should be denied CDM help for a particular type of project that was approved for another.

There may be two practical approaches. The CDM could be launched allowing a wide range of potential projects, and this range could be narrowed as experience was gained. Alternatively, the process could start with a narrow list, which could be widened in the light of experience gained. In either case, it should be made clear that there is no expectation that the CDM will be born fully formed; the scheme will in all likelihood be modified in view of both how it works and how technology develops.

• *Comparisons between different environmental consequences.* It is perfectly legitimate that international policy concerning climate change should give proper recognition to other possible environmental effects. Unfortunately, there are few, if any, technical approaches, with the possible (but disputed) exception of increasing energy efficiency, that do not have some negative environmental implications. For example, measures such as 'clean coal' burning for tackling emissions of gases associated with acid rain may have negative consequences for climate change. (The processes necessary to scrub the acidic gases from flue streams, either directly or in the combustion chamber, require energy and materials. Thus they may increase fuel use and greenhouse gas emissions per unit of energy production.) The use of renewable energy requires large areas of land (and may compete with other potential uses such as agriculture), with possible implications for river life, birds etc. Sometimes it involves the use of a variety of unpleasant chemicals. Nuclear power involves routine and accidental releases of radioactive material.

A common unit of measurement would seem to be necessary in order to determine the general environmental effect of a particular technology or installation. At present, attempts seem to be rather piecemeal, for example the inclusion of nuclear power in Britain's Climate Change Levy, on the presumed grounds that the environmental costs of nuclear power are exactly as damaging as those of fossil fuel-generated power. Some commentators have claimed that, in effect, this is merely political prejudice or cowardice dressed up as 'environmental protection' and without any rigour in its underlying reasoning.

Finding this measurement will be extremely difficult both in principle and in practice. To recognize the nature of the problem, one has only to ask how much visual intrusion from a wind farm, radioactive release from a nuclear plant or destruction of bird breeding grounds from a tidal barrage represents the same level of detriment as a tonne of carbon dioxide. Nonetheless, some common measurement would seem to be essential if the overall environmental effects of different policy proposals are to be meaningfully compared. An extensive literature exists that attempts to put financial values on economic externalities, especially emissions of atmospheric pollutants such as carbon dioxide and sulphur dioxide, but the figures arrived at vary widely. It would also seem likely that the damage associated with some environmental problems (although not greenhouse gas emissions) will vary depending on circumstances. For example, the land requirements for solar cells or the visual intrusion associated with a wind farm may be less problematic in sparsely populated countries than in more densely populated areas. The local damage associated with acid rain emissions may similarly be greater if the source of the pollution is closer to centres of population.

The European Union's ExternE work on the external costs associated with various ways of making energy is one attempt to answer these questions. It has found that the externalitics associated with coal and oil are significantly higher than those associated with gas, which in turn are higher than those connected with nuclear power or renewables. However, its calculations depend on extremely uncertain data. They also face a generic difficulty, in comparing 'soft' environmental effects such as visual intrusion and disruption of natural habitats with effects that can more easily be quantified in monetary terms. Some commentators have interpreted these observations as further support for the idea that individual countries should be allowed to define sustainable development within their own national context.

- *Flexibility*. The future is rarely a mirror of the past, as breakthroughs and gradual improvements are a constant feature of technological development and factors such as international hydrocarbon prices are notoriously volatile and unpredictable, even over short periods. It seems questionable, therefore, that any technology *per se* should be omitted from CDM positive or negative technology lists. Moreover, the exclusion of a technology from the CDM may inhibit its development in ways that could be attractive and appropriate for various developing countries.

CDM stakeholders

A CDM project would have a number of stakeholders with different sectional interests, in addition to an assumed shared concern for the environment and sustainability. At least five categories of stakeholder can be identified:

- the government of the investing country;
- business organizations in the investing country;
- the government of the host country;
- the project partner in the host country;
- NGOs.

One would also expect different members of each group to have different priorities.

The governments of investing countries would be interested in the electoral impact of a project in the context of their general policies, and would perceive it in the light of their desire to:

- demonstrate that action is being taken to protect the environment;
- reduce the (immediate) impact of a policy to mitigate climate change on domestic interests, for example the manufacturing industry and individual voters;
- avoid involvement with unpopular technologies or causes;
- ensure that domestic investment is not adversely affected.

The governments of host countries would similarly be interested in the electoral consequences of a project. Important considerations might be:

- its 'fit' with long-term development strategies;
- a desire to attract external resources to assist development;
- a desire, on the other hand, to ensure that a project does not divert resources from other, more beneficial projects or absorb too many human resources;
- a wish to avoid the impression that the host country is simply being used to allow the investing country to avoid domestic action towards limiting or reducing emissions.

For NGOs, their ability to attract new members and media attention depends on their ability to project a moral image.

The views within the business community, in both host and investing country, are likely to be more diverse. Some elements, perhaps within the fossil fuel industry, may have an interest in downplaying the threat of climate change, and thus may wish to undermine the entire Kyoto process, while others may regard the CDM as an opportunity to maintain market share in the developed world. Those wishing to sell low- or zero-carbon electricity generating technologies may regard the CDM as a boost to their export prospects.

The outcome of this pluralism is likely to be reflected in a final CDM project in which 'factual' matters are viewed through a decidedly political prism. It is difficult to avoid the impression that, to many stakeholders, mitigating climate change is by no means the most important potential outcome of an agreement under the Kyoto Protocol. To some, a result that reduces greenhouse gas emissions but involves serious costs to major corporations or to large numbers of voters is probably unacceptable. Others would not support decreases in greenhouse gas emissions that are accompanied by a major revival of the nuclear industry. Still others would oppose cuts in emissions that close off the nuclear option. Against this background, even the definition of basic terms is likely to be controversial.

The role of nuclear power in the CDM

The possible role of nuclear power in the CDM, and also within Joint Implementation, has been a topic of particular dispute, notably during the CoP-6 discussions in The Hague in 2000 and in Bonn in 2001. These discussions resulted in Annex 1 countries pledging not to use CERs generated through nuclear projects in the first compliance period.

Several studies have been carried out to estimate the relative emissions of greenhouse gases from different methods of generating electricity. Those studies commissioned by the nuclear industry tend to find that the emissions associated with nuclear power are very low, while those commissioned by anti-nuclear pressure groups assess them as higher, though still much lower than those associated with fossil fuels. Three examples, in Table 7.2, illustrate this point.

Table 7.2 Emissions of greenhouse gases from power production (grammes carbon equivalent per kWh output)

Study	Coal	Oil	Gas	Nuclear	Hydro	Photo-voltaic	Wind
Central Research Institute of the Electric Power Industry, Japan [a]	270	200	178	2.4	4.8	16	9.5
WWF [b]	281	n.a.	112	9.6	9.0	27	5.4
ETSU [c]	282	224	160	1.2	n.a.	n.a.	n.a.

[a] TEPCO (2000).
[b] WWF (2000).
[c] ETSU (1995).

Nonetheless, it is a simple statement of fact that the use of nuclear power in preference to fossil fuels (without effective sequestration) would reduce emissions of carbon dioxide, and perhaps of other greenhouse gases such as methane and nitrous oxide. The same applies to renewable technologies, to fossil fuels with sequestration of the carbon dioxide waste stream (for example by capturing the carbon dioxide and storing it in deep aquifers or as solid carbonate) and perhaps to measures for improving energy efficiency. Were reductions in emissions of greenhouse gases the sole aim of the CDM, it would be very difficult to construct an argument against including all of these technologies in the process. One would then expect the most cost-effective to be taken up in any particular circumstance.

There seems to be widespread agreement that the CDM will be a minor part of the global strategy for reducing greenhouse gas emissions, at least at first. This impression was reinforced by the concessions on carbon sinks granted at the Bonn discussions in 2001, which in effect made it easier for developed countries to meet their emission reduction commitments. However, several commentators from both the pro-nuclear and anti-nuclear viewpoints believe that the inclusion of nuclear power in the CDM would probably result in more nuclear power stations being built (although others disagree). The Chinese government, for example, has openly said as much. In the view of the nuclear industry, because nuclear power projects would be real, quantifiable, verifiable and additional, they would be ideal candidates for the CDM.

Article 12 of the Kyoto Protocol identifies three specific goals for the CDM:

- to assist in the achievement of sustainable development;
- to contribute to the attainment of the environmental goals of the UN Framework Convention on Climate Change (UNFCCC), which arose from the Rio Convention of 1992;
- to assist Annex 1 parties (the developed countries) in complying with their emission reduction commitments.

In effect, the incentive for developing countries to become involved in CDM projects, apart from improving the effectiveness of the global programme for combating climate change, is the attraction of foreign investment for aiding economic development.

The dispute over whether there should be restrictions on the type of project that might be considered for CDM (and JI) support and, if so, whether nuclear power should be included or excluded has focused largely on different concepts of sustainable development. To some, there are no inherent features of nuclear technology that would prevent it from contributing to sustainable development, and so the flexibility to expand it in the medium to long term should be preserved. To others, the drawbacks associated with large-scale power plants should preclude them from ever being regarded as 'sustainable'. Thus, the CDM should be reserved for renewable energy and energy efficiency projects. To still others, the particular features of nuclear power, especially its association with radioactive materials, make its use unsustainable.

Within these positions there are variations. Some of those who support nuclear power's place in the CDM (and outside it) argue that the technology in its present form is quite capable of contributing to sustainable development. The new generation of simpler nuclear designs, such as the AP600 and the AP1000, and more radical proposals such as the South African Pebble Bed Modular Reactor could improve matters further, assuming that they could be demonstrated on a commercial scale. In contrast, some opponents believe that nuclear power by nature cannot be sustainable and that no conceivable technical developments could change this position. It would follow that resources should not be wasted on nuclear R&D but should be directed towards renewable technologies and improved energy efficiency.

Other commentators accept that current nuclear technology would present major problems for some non-Annex 1 countries. However, they hold that there is no reason to believe that nuclear power, unlike other technologies, cannot evolve to overcome those problems or that nuclear power would necessarily be inappropriate for all non-Annex 1 countries today. In their view, banning nuclear power from the CDM would make no sense, and consideration ought to be given to allowing nuclear power to qualify for CDM support in some countries but not in others. Blanket bans on any technology would be too prescriptive, given that different countries have different needs and different starting points and that technology can develop rapidly.

As discussed later, a central problem is the lack of a generally accepted and meaningful definition of 'sustainable development'. The meaning of the phrase tends to change depending on the prejudices of the user. It is therefore not surprising that some NGOs, developed and developing countries, have argued that because candidates for the CDM such as forestation and nuclear power cannot meet their vision of sustainable development, they should be excluded. According to this view, the investor and host share responsibility for ensuring that development is sustainable, and any attempt to 'dump' in developing countries technologies that cannot be employed in the country of origin would be neo-imperialist in nature. (In the event, forestation, but not avoided defor-estation, was accepted as a suitable CDM technology in Bonn; nuclear power was not accepted.)

In this latter viewpoint, the CDM is often conceived as a mechanism to catalyse the introduction of renewable technologies in non-Annex 1 countries while these technologies are becoming established in Annex 1 countries. Its benefit will be felt not so much in the individual projects that are promoted but from the building of capacity in both developing and developed countries. Whether certain technologies should be included or excluded in the CDM takes on a symbolic importance: the CDM suggests which technological paths should be followed, and does not directly serve to reduce emissions signifi-cantly. The exclusion of nuclear power from the CDM might create a sentiment that would result in the cancellation of nuclear projects that might otherwise have gone ahead in developing and, quite possibly, developed countries.

Another area of disagreement concerns who decides about sustainable devel-opment paths for particular countries. Several states, including Japan, China and India, have argued that developing countries can decide for themselves which technologies will contribute best towards their sustainable development. In Oc-tober 2001, for example, the Chinese delegation to the International Atomic Energy Agency general conference argued in favour of the role of nuclear power in a sustainable development path, citing the Chinese government's guideline for 'developing nuclear power appropriately' in the latest five-year plan of national economic development. In this view, any attempt by the rich world to dictate which technologies could qualify for sustainable development in any particular country would in effect be neo-imperialism.

One further point should be noted. The Clean Development Mechanism at present refers only to the first compliance period, from 2008 to 2012; there are no agreed targets beyond this time. Even if nuclear power is included in the CDM, it is unlikely that any nuclear power plant that was not already planned or under construction could be completed under the CDM, and gen-erate CERs, by 2012, owing to the combined length of the planning and construction phases. Thus, the only direct significance of the exclusion or inclusion of nuclear power is symbolic: it will indicate the prevalent global attitude to the role of nuclear power in sustainable development.

However, the longer term cannot be ignored. If current projections of cli-mate change prove to be accurate, then subsequent compliance periods will be necessary, and the issue of inclusion or exclusion of nuclear projects will be of more consequence. When the compliance periods are set – the next is expected to be agreed by 2005 – the question will arise as to whether projects instituted under the CDM before the first compliance period should be eligible for credits in subsequent periods.

The deal reached at Bonn in 2001, albeit without US participation, appeared to breathe new life into the Kyoto Protocol, which had looked in poor health beforehand. However, it is quite possible that even if the modified Kyoto Protocol itself should fail, the Kyoto mechanisms might survive to become an integral part of its successor.

Nuclear power and the CDM: generic and non-nuclear issues

It is not within the scope of this paper to consider in detail the theoretical drawbacks to the CDM, although they involve some issues generally applicable to the Kyoto mechanisms:

- the determination of the level of emission reduction associated with a particular project;
- the verification of the success of the project against initial criteria;
- big incentives for participants to cheat;
- the size and management of administrative and transaction costs;
- the creation of an equitable governance structure.

This section will concentrate on the first issue, known as 'additionality', which is especially thorny. There can be no justification for the CDM if it simply rewards projects that would have taken place anyway. In that case, the CDM would serve merely to reduce the emission reductions required of the developed world. Projects carried out under the CDM must lead to a reduction that would not otherwise have occurred in the level of emissions.

Additionality is of two kinds: emission additionality and financial additionality. The emission additionality requirement means in effect that it must be demonstrated not only that the project in question reduces emissions in comparison with the likely alternative but also that the project would not have been carried out were it not for the CDM. Financial additionality has arisen because of fears among some developing countries, as suggested previously, that current aid budgets (for environmental or other projects) might be diverted towards the CDM. This might lead to a reduction in greenhouse gas emissions against current projections, but would not represent any additional investment in the host country – there would be emission additionality without financial additionality. This point, clearly, is not specific to potential nuclear projects.

The additionality requirements lead to a basic paradox. The whole aim of the CDM is to ensure that action to mitigate climate change is carried out in a cost-effective manner. However, those projects closest to being economically viable are those most likely to be carried out even in the absence of the CDM. It follows that additionality can most easily be demonstrated for those projects that are unlikely to be economically viable in the absence of certified emission reductions. But the CDM is designed to promote the more economic projects.

It is clear that establishing additionality, whether emission or financial, is likely to be quite difficult. Probably it will never be possible to be certain that a particular project had been carried out, or made more efficient, only because it would generate CERs. It may also be difficult to demonstrate that the investment had not already been earmarked for another development project in the host country, although if CDM money is coming mainly from the private sector it is less likely to replace government aid for other projects. For example, assume that a particular developing country has plans to build a series of nuclear stations. The prospect of a CDM including nuclear power would surely tempt the country publicly to cancel its nuclear plans. Entirely new plans, involving the same number of reactors but in slightly different locations, might then be published, with the proviso that these plants would be built only with the aid of CDM credits. In the absence of these credits, it might be claimed, alternative capacity fired by coal would be installed. It would presumably be the task of the CDM executive board to judge the veracity of this claim, but this job might be fiendishly complex and inevitably political.

Nevertheless, as noted, there seems to be agreement among some pro- and anti-nuclear activists that the inclusion of nuclear power in the CDM would probably result in more nuclear stations being built in developing countries than were it to be excluded. Greenpeace calculates that at a CER value of $10–30 per tonne of carbon, the inclusion of nuclear power within the CDM could reduce effective capital costs for nuclear stations by between 10 and 40 per cent.[1]

In those countries where a major expansion of electricity capacity powered by coal is projected – India and China are obvious examples, with a new coal-fired power station being commissioned every two weeks in China – the installation of a nuclear project supported by the CDM would therefore be likely to result in real reductions in emissions. (It remains possible that the other potential disadvantages of nuclear power may rule it out of consideration for the CDM if they are regarded as more serious than the effects of greenhouse gas emissions.)

---

[1] Greenpeace (2000).

Specific difficulties with the CDM

There are other potential difficulties that are specific to the CDM. They include:

* fears among developing countries that they might lose control of projects;
* fear of competition between different aid programmes, ultimately resulting in a reduction of official development assistance that might be more in line with the host country's own priorities (an issue closely connected to financial additionality);
* concern that the CDM could exhaust the cheapest options for emission reduction ('low-hanging fruit'), leaving more expensive options to the future, when the developing country had enjoyed sufficient economic growth to accept reduction targets of its own;
* fears among some developing countries, notably those in Africa and the Pacific, that the CDM might perpetuate inequalities in aid received by different regions of the developing world.

Large-scale projects

Some of the advantages and disadvantages of including nuclear power in the CDM are shared by other large-scale power projects, such as hydropower, large-scale gas-fired plants or large-scale 'clean' coal stations with sequestration of carbon dioxide emissions. On the positive side, the emission additionality of a large-scale power project might be relatively easy to determine. If, say, it were deemed likely that in the absence of the CDM, a large-scale coal-fired plant would be built, then the emissions expected from that plant could be calculated relatively simply, by reference to standard emission tables. The emissions associated with the CDM-aided alternative could also be calculated, taking into account any emissions associated with the full fuel cycle. In at least some cases one could be confident that the CDM had resulted in lower emissions than would otherwise have been the case.

Further, the relative transaction costs of CDM projects would be lower than equivalent greenhouse gas emission savings from a large number of small-scale projects, each of which would have to be validated, subjected to additionality calculations and monitored appropriately. This would probably make them more attractive, particularly to large investors. However, it looks likely that in the absence of smaller designs of nuclear power station than are now available, nuclear projects will be concentrated in countries which have at least local electricity grids or which are developing them.

This gives rise to three questions. First, might the presence of large-scale options drive out support for smaller renewable energy and energy efficiency schemes within the CDM? The alternative view is that the CDM is not a zero-sum game, and there is no reason why a credited nuclear or other large power scheme should be in competition with renewable energy projects: they would probably be attractive to different potential investors, and to different potential host countries.

As noted earlier, a second, related concern is that large-scale CDM-backed power projects might accumulate in those countries, the 'emerging economies', that already have grid systems, to the detriment of investment in the least developed countries. This concern applies to the whole CDM concept. Only a tiny minority of CDM-style projects under the AIJ pilot phase was set in sub-Saharan African countries and other least developed countries.

Third, might the inclusion of large power projects in the CDM accelerate the installation of supergrids in developing countries? Some commentators argue that centralized electricity supply systems, with national or regional supergrids, are in themselves unsustainable and cause intolerable environmental damage in the areas near the plants themselves. Others hold that centralized grids offer considerable development benefits, for example in terms of security of power supply. They argue too that nuclear projects could be attractive only in countries with a grid or a partial grid. There they would displace other large-scale power projects, in all likelihood powered by fossil fuels, particularly coal.

Nuclear power and sustainable development

Sadly, there is no single accepted definition of sustainability – none is offered in either the Rio Convention or the Kyoto Protocol. The Brundtland Commission in 1987 defined sustainable development as 'development which provides for the needs of this generation without compromising the ability of future generations to meet theirs'.

Discussion in recent years has focused on the concept of a 'sustained increase in *per capita* well being',[2] coupled with ensuring that environmental systems are not degraded in the meantime. However, this is still too vague to be of detailed use in determining the sustainability of any particular technique or technology, as a result of which opinions can differ sharply.

A more detailed consideration of the concept yields a number of elements related to well-being:

---

[2] Pezzey (1989).

- economic – a desire not to consume unnecessarily economic resources that could be used for other social goods, for example health care;
- resources – not depleting limited natural resources, especially if they have other constructive uses;
- environmental – either preventing any environmental degradation or allowing environmental degradation only if it is outweighed by other benefits;
- human and social – poverty eradication, health, education, access to leisure.

Sustainable development may therefore be conceived of as development in which the sum of well-being grows, perhaps with the proviso that environmental well-being must not decline except in the short term. By definition, of course, sustainable development must be not only sustainable – it must also promote development. Many commentators have stressed that access to affordable energy supplies, and perhaps especially to affordable electricity supplies, is as much a requirement of sustainable development as is protection of the environment, especially in the view of developing countries. It can also be argued that a sustainable development programme would include sizeable resources for research and development, with a view to solving or avoiding the unsustainable elements of currently available technology. Sustainability is a matter of protecting the future, but also of ensuring a proper allocation of well-being now. Within the concept, then, is an element of capacity building, of transferring technology and assets from developed to developing countries.

As might be expected in the absence of a clear definition, there are many different concepts of sustainability. Some commentators regard it as an absolute concept. Thus Greenpeace, in calling for a 'positive list' of allowable technologies for the CDM before the CoP in The Hague, wanted candidates that are '100 per cent emission free and 100 per cent sustainable'.[3] Others view sustainability, almost by definition, as a more relative idea, with various possibly contradictory factors to be balanced against each other; they would not speak about something as being '100 per cent sustainable'. The former attitude would lend itself more readily to a positive or negative list of candidate technologies for CDM accreditation, and the latter would imply more of a case-by-case approach, with more responsibility being vested in the CDM executive board.

---

[3] Greenpeace (2000).

The nuclear dispute

The dispute over whether nuclear power should or should not be a candidate
for inclusion within a sustainable development framework ranges over the
whole of the nuclear 'debate'. Important areas include the following:

- *Economic and resource issues.* Opponents of nuclear power argue that its
  economics are so poor in comparison to the alternatives that resources
  should not be wasted when there is sufficient potential in a combination of
  renewables and energy efficiency to guarantee major emission reductions.
  Supporters of nuclear power argue that newer generations of nuclear design will have much better economics and that including nuclear power in
  the CDM would simply allow it to compete fairly against the alternatives.
  Nuclear projects would be built only if their economics were favourable in
  comparison to alternatives. Furthermore, nuclear projects occupy far less
  land or water area than renewables. Underlying this view is a scepticism
  that energy efficiency and renewables could fulfil more than a modest proportion of the expected growth in electricity demand, caused, for example,
  by attempts to bring electricity to the one-third of the world's six billion
  people who have no access to it at present. In this context, the choice of
  fuel in at least some cases would be between nuclear power and coal.
- *Environmental and social issues.* The argument here includes issues about
  plant safety (especially in countries that may not have strong political institutions), waste management, transportation of nuclear materials and nuclear
  weapons proliferation. These are covered in detail in other chapters. Perhaps a fundamental difference is that the proponents of nuclear power tend
  to regard climate change as the dominant environmental and human problem facing decision-makers in the next decades. They think that the
  environmental disadvantages of nuclear power are generally exaggerated.
  Opponents, on the other hand, argue that other possible environmental hazards, notably those associated with radiation, should be accorded equal
  importance.
- *Signals about 'acceptability'.* Given the period of time it takes to license
  and build a large power station, few, if any, entirely new nuclear projects
  financed with CDM support could generate electricity by the end of the
  first compliance period, in 2012. However, the decision not to include
  nuclear power in principle for the first compliance period was important, as
  it undoubtedly sent a signal about the way the international community regarded nuclear power. To opponents of nuclear power, its eventual effective
  exclusion from the CDM implied that it was somehow not a suitable

technology, not just for developing countries but for developed ones too. This message was an appropriate one, given their perceptions of a failed and expensive technology that has diverted major research and other resources away from finding sustainable solutions to development challenges. However, there was widespread suspicion among proponents of nuclear technology, and others, that the matter was simply a political 'fix' designed to achieve some kind of agreement from the process. They felt it irrational to impose blanket bans on any particular technology, since no project could proceed unless it had financial backers and was acceptable to the host country. This dispute will re-emerge in the discussion about the next compliance period, expected to conclude in about 2005.

Summary

The Clean Development Mechanism will not spring forth fully formed. In the light of early experience, much flexibility and fine-tuning will be necessary in order to steer a course between CDM projects in which criteria are so difficult to meet that none take place or so easy to meet that developed countries can offset most of their emission reduction requirements with CDM schemes that are of dubious effectiveness.

Either of two models of the executive board of the CDM may therefore emerge. In the first, it would have very wide discretion over which projects would be accepted in particular circumstances. No technology would be ruled out as a matter of policy, but the possible environmental disadvantages of a particular scheme would have to be taken into account in assessing its eligibility. This model would allow considerable flexibility over which types of project may be accepted, for example as understanding of the effects of different technologies increases. As the board would have considerable powers in the area of policy-making, the potential for discord, for example if one developing country were regarded as a suitable host for a CDM-aided nuclear power project but another was not, could be considerable.

The second model would see the executive board's discretion fettered by policy decisions taken by Conferences of Parties to the UNFCCC, for example in formulating positive or negative lists. The board's job would be a narrower and more technical one: assessing the emissions of greenhouse gases that would have occurred in the absence of the project, assessing the emissions that would occur if the project were to go ahead, monitoring the whole process and awarding CERs. Although this approach may be more equitable in some senses, and would locate policy issues with the CoPs rather than the board, it would clearly be less flexible in its response both to different national circumstances and to changes in technologies and perceptions over time. The resolution of this issue is likely to be quite important in the practical development of the CDM.

The CoP held in Bonn 2001 had two important consequences for the issue of whether or not to include nuclear power in the CDM:

*   an agreement by Annex 1 countries to refrain from using CERs generated through nuclear projects in other countries (developed or developing);
*   a dilution of the requirements for emission reductions through the inclusion of carbon sinks, with the result that compliance with Kyoto Protocol commitments would be easier to achieve without a large number of CDM projects.

Even though there is now little chance of including a nuclear project in the first compliance period (ending in 2012), nuclear exclusion or inclusion will remain an important issue, especially through negotiations, expected to conclude in 2005, about the second compliance period. Assuming that more cost-effective nuclear technology has been demonstrated by 2005 or soon afterwards, new nuclear construction could well be an attractive option for reducing greenhouse gas emissions in the second period and beyond, especially if the new technology can be built more quickly than present reactors. Furthermore, the inclusion or exclusion of nuclear power in the CDM will send clear signals about the international community's view of nuclear power as a potential contributor to sustainable development. This may in turn have effects on the number of reactors being built in the industrialized world and in economies in transition.

Nuclear power's claims to be included in a sustainable development framework are hotly disputed. Its supporters see as a clear attraction the facts that nuclear power does not depend on a source of fuel required for other uses, that its fuel is not likely to run short and that it is not concentrated in a small number of countries. Its waste products do not contribute significantly to climate change, and can be managed safely. A nuclear station occupies far less land or water than would a renewable energy project of an equivalent output, and the nuclear power stream does not depend on local weather conditions. In its opponents' view, the waste implications of nuclear power are far from resolved, and are perhaps not resolvable in principle. The association of nuclear power with nuclear weapons, and the implications of the technology in times of terrorism or a breakdown of the host society, are unacceptable.

These issues are unlikely to be resolved quickly, if at all. As far as the CDM is concerned, perhaps the central question is who should have the power to decide. Should the potential host country, together with the potential investor, be allowed to choose its own definition of and strategy for sustainable development, without facing restrictions from the international community by way of exclusion of particular technologies from the CDM? Or are the international implications of some technologies, notably nuclear power, so great that there should be a list of prohibited – or allowed – technologies eligible for financial support under the CDM, arrived at by (inevitably political) trading at CoP level?

References

ETSU (1995), *Full Fuel Cycle Atmospheric Emissions and Global Warming Impacts from UK. Electricity Generation*, ETSU-R-88. London: HMSO.

Greenpeace (2000), *The Clean Development Mechanism – Used by Renewable Energy or Abused by Coal and Nuclear*, briefing paper. Amsterdam: Greenpeace.

Pezzey, J. (1989), *Economic Analysis of Sustainable Growth and Sustainable Development*, Working Paper 15, Environment Department. Washington, DC: World Bank.

TEPCO (2000), 'Carbon Analysis for the Life Cycle of Power Generation Systems', *E7 Observer*, No. 25, Special Issue, 'Nuclear power in the scope of sustainable development', 2000.

UNFCCC (2002), *Activities Implemented Jointly, http://unfccc.int/program/uij/uijproj.html*.

WWF (2000), *Climate Change and Nuclear Power*. Gland, Switzerland: WWF.

8 Recommendations and conclusions

Recommended actions

The nuclear option will always remain 'open', in the sense that the technology is understood. Also, records can be maintained even if no more stations are built and existing ones end operation over, say, a 50-year period. To restart the nuclear industry, however, would be a major and lengthy undertaking, and problems in the energy field can emerge rapidly. The uncertainties, and the size of the challenges, associated with the issue of energy and the environment over the next decades are large. Thus, it can be argued that action should be taken today to ensure that nuclear power is available as a practical option.

The extent to which action should be taken is a matter for debate. The answer will depend on factors such as perceptions of the size of the problems and of the extent to which nuclear technology can evolve and matters of politics and values. The illogicality of the argument that of all the energy sources, only nuclear power will fail to evolve and respond to changing conditions is clear. But the extent to which that evolution can be carried out successfully will depend largely on actions taken or not taken by the nuclear industry and by governments.

The following lists of actions, which include several initiatives already in train, are not intended to be imperative. Some – especially those associated with demonstrating new reactor technologies, managing radioactive waste and developing more democratic decision-making in complex technical fields – may be more important than others if investment in new nuclear stations is to be made. Others, such as ensuring a flow of appropriately skilled individuals, will be necessary whether the nuclear industry flourishes or dies.

In view of the timescales involved, serious consideration must be given to what additional actions, if any, are required now and in the near future if the nuclear option is practically to be kept open for the year 2020 or even 2050. The following lists are classified into two categories – areas in which action can and should be taken now and more strategic issues which should be addressed over the next decade or so. (Of course, investigation may reveal that some problems are unsolvable.)

Table 8.1 Public perceptions and decision-making

Actions required of the nuclear industry	*Actions required of governments*
Short term Ensure that management becomes genuinely open, with an assumption that information will be made available to Interested parties and that legitimate issues such as commercial confidentiality and security are not used as an excuse for secrecy. *Medium term* Become properly involved in appropriate new decision-making structures and procedures.	*Short term* Ensure progress in developing measures to determine the underlying attitudes and needs of various groups of the 'public' towards issues such as what would constitute an acceptable waste management approach or an appropriate role for government in supporting the building of demonstration plants. Ensure that membership of regulatory and advisory bodies is open to a wide range of expertise from different interested parties. *Medium term* Investigate and support trials to develop and evaluate new decision-making and consensus-building procedures, as required by the social and political conditions in each country. Create structures within government whereby decisions with very long time horizons, such as radioactive waste management policy, can be taken within the inevitably short-term approach of politics. Exercise political leadership over eventual decisions, including playing a role as provider of balanced information and promoter of understanding of complex issues.

Table 8.2 Economics

Actions required of the nuclear industry	Actions required of governments
Short term	*Short term*
Provide evidence as soon as possible that a number of new reactor designs can deliver substantial economic improvements in the competitive position *vis-à-vis* other fuels, especially in terms of lower initial investment costs, shorter construction times and more reliable performance in their early lifetime than has been the case with many existing plants.	Introduce appropriate regimes, on an international basis, whereby the polluter pays an economic penalty for emissions of greenhouse gases and for other environmental effects.
	Medium term
Medium term	Promote transparency about all subsidies, overt and hidden, offered to various energy sources, with a view to making the assessment of relative costs more rigorous.
Clarification and resolution of the financial implications of the back-end of the nuclear cycle, such as waste management and plant decommissioning.	Create a framework to finance the management of long-term liabilities such as waste and decommissioning.
For some markets, the development of smaller, modular nuclear units.	Create a regulatory regime that is more stable and predictable, more rapidly responsive and, where possible, simpler, in order to minimize unnecessary and costly delays.
	Encourage the international harmonization of regulation, to minimize the often costly need to customize international designs to the requirements of each national regulatory body.
	Ensure a supply of suitably qualified personnel, including broad-based engineering graduates and personnel for regulatory bodies.

Table 8.3 Waste, reprocessing and proliferation

Actions required of the nuclear industry	Actions required of governments
Short term Ensure that funding will be available to discharge liabilities when they become due, for example by establishing secure and segregated funds. *Medium term* Contribute to continuous research and development of various waste management options, including the possibility of longer-term storage – on- or off-site, on- or near-surface – against the possibility that deep disposal proves to be difficult to pursue in the immediate future or unacceptable in principle. Consider management approaches for separated plutonium, including immobilization should its utilization as a fuel turn out to be unattractive. Investigate more proliferation-resistant fuel cycles and procedures.	*Short term* Step up surveillance and security around stored nuclear materials, perhaps especially in the former Soviet Union. *Medium term* Ensure that research is undertaken not just on technical issues (such as partition and transmutation) and geology (for example site characterization) but also on social issues concerning site selection and community involvement in decision-making. Explore the regulatory and legal position on international repositories for possible use by countries with few nuclear facilities and/or limited expertise.

Table 8.4 Safety

Actions required of the nuclear industry	Actions required of governments
Short term Increase levels of interplant and international collaboration. Further develop procedures to ensure that a sound safety culture is established and maintained in all operations, including adequate staffing and financing. Make sure that the consequences of unsafe operation are appreciated by all participants. *Medium term* Monitor levels of morale among staff in the face of uncertainty about their future and take action where appropriate.	*Short term* Create realistic guidelines for regulators and guarantee their continued independence. Ensure sufficient funds are available for regulatory bodies, including training, employment and resources. *Medium term* Make further efforts to improve understanding of the health effects of low-level radiation.

Table 8.5 Research, development and commercialization

Actions required of the nuclear industry	Actions required of governments
Short term Provide research and development resources to bring near-to-commercialization designs forward, with a view to widening the range of products available for different structures of electricity supply markets and to identifying the most promising options.	*Short term* Ensure that demonstration plants of those technologies close to commercial realization can be built. Widen the use of instruments such as tax breaks, to encourage commercial companies to increase R&D activity over a range of technological areas. Carry out a thorough assessment of the international potential of various longer-term options (nuclear and non-nuclear) requiring research and identify key areas for nuclear research, for example new reactors, new sources of fuel, partition and transmutation of waste.
Medium term In partnership with governments, extend the degree of international collaboration on new designs, as part of the increasing internationalization of the plant construction industry.	
	Medium term Ensure that the total resources going into global energy R, D & C are appropriate to the scale of predicted global problems over the next 50 years or so. Within an international context, develop a coherent portfolio approach to R&D that provides appropriate funds for the various available options.

Conclusions

The aim of this project has not been to come to judgments as to whether nuclear power should or should not be a major component, a minor component or no component at all in providing future energy supplies. Rather, it has been to expound and develop, from an uncommitted standpoint, the arguments used by proponents and opponents of the technology. Nonetheless, we do feel that it is appropriate to highlight some themes that have emerged.

There are enormous variations in the role that nuclear power has played and plays today among the regions and countries of the world, and it is thus a mistake to attempt to generalize too much. However, it is fair to say that 'society' in many industrialized countries has given nuclear power one chance already. In the 1970s and 1980s, largely as a response to the oil crises of 1973 and 1979–80, many industrialized countries turned to nuclear power as an alternative to increasing their reliance on fossil fuels in an uncertain and apparently hostile world.

The outcomes of that opportunity are disputed between supporters and critics of the technology. Its supporters conclude that nuclear power has largely fulfilled its early promise: it now generates about one-sixth of the world's electricity, having been the fastest-growing of the major energy sources in proportional terms in the 1970s, 1980s and 1990s. It does so safely – it is among the safest of the major energy sources, according to some studies – and without emitting significant quantities of greenhouse gases. Its opponents hold that nuclear power has fallen far short of its early claims and indeed of public expectations in terms of economics, the failure to find an acceptable waste management approach, the potential for major accidents and terrorist attacks and also the highhanded way the industry has behaved towards society. A 'second chance' should be contemplated only in the most extreme of circumstances, if at all.

The reality, we suspect, lies between the extremes. Judged with reference to the early claim that nuclear electricity would be 'too cheap to meter' – by no means the general view at the time, to be fair – nuclear power has clearly not delivered, its projected economics being perhaps its most serious perceived drawback in competitive markets. Significant improvements were made in plant performance in the 1990s, resulting in a much more vigorous market for existing reactors, but in many cases acceptable reliability was achieved only after several years of operation. The arrogant attitude of the industry and of some of its supporters in government has been a factor in the public's loss of confidence in a number of developed countries, and no comprehensive long-term waste management route has yet been demonstrated.

Yet the safety record has been impressive: only one accident has had demonstrable, though significant, off-site health consequences. The use of nuclear power has also prevented the emission of large quantities of gases associated with problems such as climate change and acid rain, although its critics argue that had other low-carbon technologies been given the same support as nuclear power, they would have delivered equivalent or perhaps greater emission reductions. The existence of nuclear power as an alternative to fossil fuels for power generation may have had some downward effect on the prices of those fuels, though it is difficult to quantify any such effect.

The extent to which nuclear power will appear attractive in future will depend on perceptions of two main factors – the 'environment' in which it is operating and its own intrinsic features. The external environment will include elements such as:

- global energy demand (itself dependent on factors such as population growth, levels of global economic activity and the success of energy efficiency measures);

- the availability of fossil fuels (affected by factors such as the size of the global reserves and the politics of international trade);
- the degree to which renewables fulfil the predictions of their supporters;
- the severity of climate change;
- the progress in bringing to commercialization the large-scale sequestration of carbon dioxide from the use of fossil fuels;
- the structure of electricity supply systems (involving the degree and type of regulation and the extent to which systems rely on embedded or centralized generation);
- the political ideology of particular governing parties;
- the state of public opinion towards issues such as radioactive waste management, safety, the security of electricity supplies from other sources, climate change etc.

These factors, with the partial exception of the last one, are mainly outside the control of the nuclear industry itself. In a future of, for example, energy shortages, disappointing performance by renewables and acute fears about climate change, nuclear power would presumably look more attractive than it would in a future of limited energy demand, a flourishing renewables industry and perceptions that climate change is manageable.

As noted earlier, the nuclear industry itself could, in principle, take a number of steps to make itself more attractive. But whether these steps – say 'establishing an acceptable waste management approach' or 'developing smaller and cheaper reactors' – would be possible in reality is for the moment an open question.

Suppose that acceptable technical solutions, at a reasonable cost, can be developed for the major areas of concern about nuclear power. It might nonetheless be very difficult in practice to develop those solutions. A number of log-jams will first have to be overcome. For example:

- companies might not be prepared to put in the research, development and commercialization effort necessary to demonstrate cheaper and safer nuclear designs without a reasonable prospect that those designs will find a market, but a market may not emerge until the designs are ready;
- the development of novel waste management techniques such as partition and transmutation may make sense only if there is an expanding nuclear industry, but expansion may be impossible without new ways of managing waste;
- public perception may remain inimical to nuclear power unless there is a major programme of new reactors (that would challenge the 'fear of the

unknown' and perhaps perceptions of risk of accidents), but a programme may be impossible until public fears are overcome;
- the construction of further nuclear plants would require an adequate flow of skilled personnel capable of designing, building and running power stations and other nuclear facilities and of staffing regulatory bodies. But these individuals, and the infrastructure of courses and other training for supporting them, probably would not be attracted into the field if they perceived that there might not be any work for them.

Similar problems may be encountered with respect to renewables and carbon dioxide sequestration, and perhaps even in the case of demand-side technologies.

To discover whether solutions to the major areas of difficulty are feasible, governments, either alone or in collaboration, will have to act now or very soon and provide the initial stimulus for progress. Certainly, governments will have to create the circumstances in which there is a sufficient flow of suitably qualified individuals for staffing research and regulatory bodies – this is true whether or not new nuclear power stations are built. It is highly likely that governments will also have to ensure that plant operators put aside sufficient funds for dealing with waste management and decommissioning in the long term.

Perhaps the most difficult issue concerning nuclear power is the construction of demonstration plants – either those that have been or are about to be licensed but have not yet been constructed (Generation III) or more speculative and long-term concepts (Generation IV). Eventually, these plants will be necessary in order to reveal whether the new designs of reactor referred to in Chapter 6, on R, D & C, will be sufficiently attractive to gain commercial support. As noted earlier, demonstration facilities might also be required in other technical fields, such as radioactive waste management, large-scale renewables, carbon dioxide sequestration or more innovative demand-side technologies. But if private companies are unwilling or unable to build demonstration facilities, the financial risk being too great, we conclude that governments ought to take steps to ensure that demonstration plants are built. These initiatives could include the awarding of direct grants. As the energy industries are increasingly international, these demonstration facilities might well involve collaboration among a number of countries and companies, thus reducing the commitment of any one participant.

If society is in effect to give nuclear power a 'second chance', the lessons of the 1970s and 1980s will have to be learnt and acted upon. In particular, procedures that are more open and democratic, by which the public at large is able to influence the course of development, must be developed and followed, especially in those countries in which more 'traditional' approaches to decision-making have been ineffective.

Early work in this area has been carried out, with some degree of success. As the industry has become more commercialized in the countries in which it had its origins and has lost its favoured position with governments, so it seems to have lost some of its early arrogance. This trend must continue if the feeling, still prevalent in some circles, that nuclear power is something imposed upon rather than part of society is to be overcome. This feeling, not directed only at the nuclear industry by any means, could act as a strong brake to nuclear development in societies experiencing the 'decline of deference', even if the other arguments for nuclear investment are strong.

In summary, two key issues face the nuclear power industry, governments and all others who can see a case for keeping the nuclear option open:

- what needs to be done to enable decision-makers to assess whether concepts promising at the R&D stage can be successfully commercialized;
- what needs to be done to develop decision-making structures that can manage complex technical, economic, environmental and social issues such as nuclear power (and many others).

In the immediate future, it looks likely that the 'centre of gravity' of nuclear activity will continue to move away from North America and western Europe towards South and East Asia. Before long, however, a new understanding among the people, governments and nuclear industries may be needed. This understanding should open the way for a proper international appraisal of whether, and in what circumstances, nuclear energy might make a positive contribution to meeting the energy and environmental challenges that the world faces in the twenty-first century.

Glossary

Actinides	Heavy and often long-lived radioactive elements, of which the most important are uranium, neptunium, plutonium, americium and curium.
ADS	Accelerator-driven systems.
AIJ	Activities Implemented Jointly, the pilot project for JI and the CDM introduced at the UNFCCC CoP-1 in Berlin, 1995. (*See* JI).
ALARP	The principle that any dose of radiation sustained by workers or the public should be 'as low as reasonably practicable'.
Alpha radiation	Particles equivalent to helium nuclei, emitted during alpha radioactive decay, especially of actinides. They cause much disruption to cells if released by material inside the body, although they cannot penetrate the skin.
AP600, AP1000	Advanced passive reactors (BNFL Westinghouse).
ATW	Accelerator-driven transmutation of waste.
Beta radiation	Particles the equivalent of fast-moving electrons, emitted during beta radioactive decay, especially of fission products. They can represent a health hazard if emitted inside or outside the body.
BSLs	Basic safety limits.
BWR	Boiling water reactor.
CANDU	Canadian deuterium uranium reactor.
CCGT	Combined cycle gas turbine.
CDM	The Clean Development Mechanism of the Kyoto Protocol.
CERs	Certified emission reductions, generated when a CDM project is carried out and can be demonstrated to have reduced or limited emissions of GHGs.
CoP	A conference of the parties to the UNFCCC.
EC	European Commission.
EPR	The Franco-German European pressurized water reactor.
FGD	Flue gas desulphurization.
Fission products	Smaller atoms created during the splitting of uranium or plutonium atoms, many of which have relatively short half-lives.

FR	Fast reactor, which can operate on plutonium or highly enriched uranium, and can be used to burn or to breed plutonium fuel.
Gamma radiation	Bursts of energy emitted during gamma radioactive decay. They can represent a health hazard if emitted inside or outside the body.
GHGs	Greenhouse gases – the most important of which are carbon dioxide and methane.
gtoe	A unit of energy, equivalent to burning one billion tonnes of oil.
GW	Gigawatt (unit of power).
Half-life	The period of time for half of a radioactive sample to decay. Isotopes with short half-lives tend to be intensely radioactive.
HALW	Highly active liquid waste.
HLW	High-level waste.
HTGR	High-temperature gas-cooled reactor.
IAEA	International Atomic Energy Agency.
IEA	International Energy Agency (a branch of the OECD).
ILW	Intermediate-level waste.
INES	International Nuclear Event Scale.
INPO	Institute of Nuclear Power Operators (United States).
INPRO	International Project on Innovative Nuclear Reactors and Fuel Cycles.
IPCC	Intergovernmental Panel on Climate Change, the main international body researching the science behind climate change, its possible consequences and measures that could be taken to adapt to it.
JI	Joint implementation, projects carried out under the Kyoto Protocol between two developed countries to reduce GHG emissions, with CERs being shared by the parties. (*See* AIJ)
kWe	Unit of installed electrical capacity.
LILW	Low- and intermediate-level wastes.
LLW	Low-level waste.
LWR	Light water reactor.
MOx	Mixed (plutonium and uranium) oxide fuel.
mSv	Millisievert, a unit of radiation dose.
NEA	Nuclear Energy Agency (part of the OECD).
NGOs	Non-governmental organizations.
NIMBY	'Not in my back yard'.

NNPT	Nuclear Non-Proliferation Treaty.
ODA	Official Development Assistance.
OECD	Organisation for Economic Cooperation and Development, a grouping of 30 developed countries with market economies.
OSPAR	Convention for the Protection of the Marine Environment of the North East Atlantic, with 15 countries and the EU as parties.
P&T	Partition and transmutation of waste or spent fuel.
PBMR	Pebble Bed Modular Reactor.
PSA	Probabilistic safety assessment.
PWR	Pressurized water reactor.
RBMK	The Soviet water-cooled, graphite-moderated reactor – the design used at Chernobyl.
RCF	Rock Characterization Facility, often an underground laboratory for examining the characteristics of rock formations.
TBq	Terabecquerel, a unit of radioactivity.
Trans-uranics	Actinides with atomic numbers higher than that of uranium, of which neptunium, plutonium and americium are the most important.
TRU	Trans-uranic waste.
TWh	Terawatt-hour, a unit of electrical energy.
UNCED	The UN Conference on Environment and Development, held in Rio de Janeiro in 1992.
UNFCCC	UN Framework Convention on Climate Change, a product of UNCED.
UNSCEAR	UN Scientific Committee on the Effects of Atomic Radiation.
VHLW	Vitrified high-level waste.
VLLW	Very low-level waste.
VVER	Soviet reactor design, similar to the PWR.
WANO	World Association of Nuclear Operators.
WIPP	Waste Isolation Pilot Plant, a deep repository for TRU waste near Carlsbad, New Mexico.

THE ROYAL INSTITUTE OF | Sustainable Development
INTERNATIONAL AFFAIRS | Programme

THE NEW ECONOMY OF OIL
Impacts on Business, Geopolitics and Society

**By John V. Mitchell
with Koji Morita,
Norman Selley and Jonathan Stern**

**Hardback £40.00 ISBN 1 85383 745 8
Paperback £16.95 ISBN 1 85383 796 2
RIIA/Earthscan Published February 2001**

The chief concern of the oil industry over the next 20 years will not be its availability, but its acceptability. Oil now faces an unprecedented series of challenges. As a major polluter, it faces competition from other energy sources – gas and renewables – and more stringent regulation and control, with higher taxes.

Faced with these pressures, the oil companies are repositioning themselves as energy industries – and oil is certain to have a diminishing share in their portfolio of fuels. The implications are enormous, given the huge current dependence on oil of so much industry and government revenue.

The book will be of importance to all those involved with oil – from industry professionals to competitors, commentators, investors, managers, politicians and regulators.

Distributed in North America by the Brookings Institution

THE ROYAL INSTITUTE OF INTERNATIONAL AFFAIRS | Sustainable Development Programme

TECHNOLOGY TRANSFER FOR RENEWABLE ENERGY
Overcoming Barriers in Developing Countries

By Gill Wilkins

Paperback £18.95 ISBN 1 85383 753 9
RIIA/Earthscan Published March 2002

This book highlights the role that renewable energy can play in achieving sustainable development. It focuses on rural areas of developing countries, looking in particular at stand-alone solar home systems and grid-connected biomass cogeneration plants. It provides a summary of the main barriers to the successful transfer of renewable energy technology, illustrated by case studies drawn from Indonesia, the Philippines, Vietnam, Thailand, the South Pacific, Kenya and India. Options for overcoming the barriers and the role of key players are presented.

The book also outlines the potential role of the Clean Development Mechanism of the Kyoto Protocol in facilitating renewable energy technology transfer in the context of climate change. This book will appeal to academics, consultants, technology manufacturers, international funding bodies, multilateral and bilateral aid agencies, policy makers and planners in developing countries.

Distributed in North America by the Brookings Institution

THE ROYAL INSTITUTE OF INTERNATIONAL AFFAIRS | Sustainable Development Programme

CLIMATE CHANGE AND POWER
Economic Instruments for European Electricity

By Christiaan Vrolijk

Hardback £35.00 ISBN 1 85383 821 7
Paperback £19.95 ISBN 185383 822 5
RIIA/Earthscan Published August 2002

The electricity sector is one of the largest carbon emitters in Europe. To control these emissions, economic instruments such as emissions trading, taxes and various voluntary agreements are increasingly being considered. However, electricity systems in Europe are diverse, appropriate policies may differ widely and similar instruments may have different effects in different countries.

This study aims first to define the various economic policy instruments being used or considered in the electricity sector, including the point of application and protection measures such as border adjustments. Second, it describes the main characteristics of the major European electricity sectors, including the different generation mixes and options and divergent policy cultures. Third, conclusions are drawn concerning the instruments likely to be developed, on both national and EU levels, their potential impact and the potential interactions of different instruments within and between countries, including issues relating to international electricity and emissions trading.

Distributed in North America by the Brookings Institution